Building Python Microservices with FastAPI

Build secure, scalable, and structured Python microservices from design concepts to infrastructure

Sherwin John C. Tragura

BIRMINGHAM—MUMBAI

Building Python Microservices with FastAPI

Copyright © 2022 Packt Publishing

Associate Group Product Manager: Pavan Ramchandani

Publishing Product Manager: Bhavya Rao

Senior Editor: Mark Dsouza

Content Development Editor: Divya Vijayan

Technical Editor: Shubham Sharma

Copy Editor: Safis Editing

Project Coordinator: Sonam Pandey

Proofreader: Safis Editing

Indexer: Tejal Daruwale Soni

Production Designer: Joshua Misquitta

Marketing Coordinators: Anamika Singh and Marylou De Mello

First published: August 2022

Production reference: 1260822

Published by Packt Publishing Ltd.

Livery Place

35 Livery Street

Birmingham

B3 2PB, UK.

ISBN 978-1-80324-596-6

www.packt.com

This book is for everyone who became my support system during those months in 2021 when I got sick and for everyone who fought and stayed strong during this pandemic.

– Sherwin John C. Tragura

Contributors

About the author

Sherwin John Calleja Tragura is a subject matter expert on Java, ASP .NET MVC, and Python applications with some background in frontend frameworks. He has managed a team of developers to build various applications related to manufacturing and fixed assets, document management, records management, POS, and inventory systems. He has a background in building **laboratory information management systems (LIMS)** and hybrid mobile applications as a consultant. He has also provided corporate Bootcamp training services since 2010 for courses on Python, Django, Flask, Jakarta EE, C#, ASP .NET MVC, JSF, Java, and some frontend frameworks. He has authored books such as *Spring MVC Blueprints* and *Spring 5 Cookbook* and a Packt video, *Modern Java Web Applications with Spring Boot 2.x.*

I want to thank my friends, Icar, Mathieu, and Abby, and my cousins, Rhonalyn, Mica, and Mila, for their time and effort when I was not well in 2021. Also, I want to express my gratitude and appreciation to the Packt team for their understanding and consideration given to me. Lastly, this book would have been impossible without the support and time of Owen, who was always there pushing me to finish the book despite the scary situation during the pandemic.

About the reviewers

Glenn Base De Paula is a product of the University of the Philippines Integrated School and is a computer science graduate of the country's most prestigious institutions, the University of the Philippines. He has 14 years of industry experience, which was mostly with the government's ICT institute and recently in the banking industry.

He uses Spring, Grails, and JavaScript for his day-to-day activities. He has developed numerous Java web applications for various projects and was also the technical team lead for several others. He currently manages a team of Java developers assigned to different projects in one of the country's most reputable banks.

He is often involved in systems analysis and design, source code review, testing, implementation, training, and mentoring. He learns about better software architecture and design in his free time.

Sathyajith Bhat is an experienced Site Reliability Engineer (SRE) with over 15 years of experience in DevOps, Site Reliability Engineering, System Architecture, Performance Tuning, Cloud Computing, and Observability. He believes in getting the most out of the tools that he works with. Sathyajith is the author of *Practical Docker with Python* and a co-author of *The CDK Book*. He loves working with various communities and has been recognized as an AWS Community Hero for his contributions to the AWS community.

I would like to thank my wife, Jyothsna, for being patient and supporting me both in my career and while working on this book.

Table of Contents

3

Investigating Dependency Injection   47

4

Building the Microservice Application   77

Part 2: Data-Centric and Communication-Focused Microservices Concerns and Issues

5

Connecting to a Relational Database   107

6

Using a Non-Relational Database   161

7

Securing the REST APIs 213

8

Creating Coroutines, Events, and Message-Driven Transactions   253

Part 3: Infrastructure-Related Issues, Numerical and Symbolic Computations, and Testing Microservices

9

Utilizing Other Advanced Features   287

10

Solving Numerical, Symbolic, and Graphical Problems   325

11

Adding Other Microservice Features   361

Preface

This book teaches you about the components of the FastAPI framework and how to apply these components with some third-party tools to build microservices applications. You will need a background in Python programming, knowledge of the principles of API development, and an understanding of the principles behind creating an enterprise-grade microservice application. This is more than a reference book: it provides some code blueprints that will help you solve real-world applications while elaborating on and demonstrating the topics of each chapter.

Who this book is for

This book is for Python web developers, advanced Python users, and backend developers using Flask or Django who want to learn how to use the FastAPI framework to implement microservices. Readers knowledgeable about REST API and microservices will also benefit from this book. Some parts of the book contain general concepts, processes, and instructions that intermediate-level developers and Python enthusiasts can relate to as well.

What this book covers

Chapter 1, Setting Up FastAPI for Starters, introduces how to create FastAPI endpoints using the core module classes and decorators and how the framework manages incoming API requests and outgoing responses.

Chapter 2, Exploring the Core Features, introduces FastAPI's asynchronous endpoints, exception handling mechanism, background processes, APIRouter for project organization, the built-in JSON encoder, and FastAPI's JSON responses.

Chapter 3, Investigating Dependency Injection, explores the **Dependency Injection (DI)** pattern utilized by FastAPI to manage instances and project structure using its `Depends()` directive and third-party extension modules.

Chapter 4, Building the Microservice Application, is on the principles and design patterns that support the building of microservices, such as decomposition, property configuration, logging, and domain modeling strategy.

Chapter 5, Connecting to a Relational Database, focuses on Python **Object Relational Mappers (ORMs)**, which can integrate seamlessly with FastAPI to persist and manage data using a PostgreSQL database.

Chapter 6, Using a Non-Relational Database, showcases the PyMongo and Motor engines, including some popular Python **Object Document Mapper (ODMs)**, which can connect FastAPI applications to a MongoDB server.

Chapter 7, Securing the REST APIs, highlights FastAPI's built-in security module classes and explores some third-party tools such as JWT, Keycloak, Okta, and Auth0 and how they are applied to implement different security schemes to secure an application.

Chapter 8, Creating Coroutines, Events, and Message-Driven Transactions, focuses on the details of the asynchronous aspect of the FastAPI, such as the use of coroutines, the asyncio environment, asynchronous background processes using Celery, asynchronous messaging using RabbitMQ and Apache Kafka, SSE, WebSocket, and asynchronous events.

Chapter 9, Utilizing Other Advanced Features, contains other features that FastAPI can provide, such as its support for different response types, the customization of middleware, request and response, the application of other JSON encoders, and the bypassing of the CORS browser policy.

Chapter 10, Solving Numerical, Symbolic, and Graphical Problems, highlights the integration of FastAPI with the numpy, pandas, matplotlib, sympy, and scipy modules to implement API services that can perform numerical and symbolic computations to solve mathematical and statistical problems.

Chapter 11, Adding Other Microservice Features, discusses other architectural concerns, such as monitoring and checking the properties of API endpoints at runtime, OpenTracing, client-side service discovery, managing repository modules, deployment, and creating monorepo architectures with Flask and Django apps.

To get the most out of this book

This book requires some experience with Python programming using Python 3.8 or 3.9, as well as some API development experience with any Python framework. Knowledge of the standards and best practices of coding Python, including some advanced topics such as creating decorators, generators, database connectivity, request-response transactions, HTTP status codes, and API endpoints, is required.

Software/hardware covered in the book	Operating system requirements
Python 3.8/3.9	Windows, macOS, or Linux
VS Code editor	Latest version of any OS
PostgreSQL 13.x	64-bit version of any OS
MongoDB 5.x	64-bit version of any OS
Mongo Compass	64-bit version of any OS
Mongo Database Tools	64-bit version of any OS
RabbitMQ	Latest version of any OS
Apache Kafka	Latest version of any OS

Software/hardware covered in the book	Operating system requirements
Spring STS	Latest version and configured to use Java 12 JDK
Docker Engine	Latest version of any OS
Jaeger	Latest version of any OS
Keycloak	Version that works with Java 12
Bootstrap 4.x	
OpenSSL	64-bit version of any OS
Google Chrome	

Open an account in Okta and Auth0 for the OpenID connect security scheme. Both prefer a company email for signing up.

If you are using the digital version of this book, we advise you to type the code yourself or access the code from the book's GitHub repository (a link is available in the next section). Doing so will help you avoid any potential errors related to the copying and pasting of code.

Each chapter has a dedicated project prototype that will describe and explain the topics. If you get lost during setup, each project has a backed-up database (`.sql` or `.zip`) and a list of modules (`requirements.txt`) to fix some issues. Run the `\i` PostgreSQL command to install the script file or use the mongorestore from the installed Mongo Database Tools to load all the database content. Also, each project has a mini-readme that gives a general description of what the prototype wants to accomplish.

Download the example code files

You can download the example code files for this book from GitHub at `https://github.com/PacktPublishing/Building-Python-Microservices-with-FastAPI`. If there's an update to the code, it will be updated in the GitHub repository.

We also have other code bundles from our rich catalog of books and videos available at `https://github.com/PacktPublishing/`. Check them out!

Download the color images

We also provide a PDF file that has color images of the screenshots and diagrams used in this book. You can download it here: `https://packt.link/ohTNw`.

Conventions used

There are a number of text conventions used throughout this book.

`Code in text`: Indicates code words in text, database table names, folder names, filenames, file extensions, pathnames, dummy URLs, user input, and Twitter handles. Here is an example: "The `delete_user()` service is a `DELETE` API method that uses a `username` path parameter to search for a login record for deletion."

A block of code is set as follows:

```
@app.delete("/ch01/login/remove/{username}")
def delete_user(username: str):
    del valid_users[username]
    return {"message": "deleted user"}
```

When we wish to draw your attention to a particular part of a code block, the relevant lines or items are set in bold:

```
@app.get("/ch01/login/")
def login(username: str, password: str):
    if valid_users.get(username) == None:
        return {"message": "user does not exist"}
    else:
        user = valid_users.get(username)
```

Any command-line input or output is written as follows:

```
pip install fastapi
pip install uvicorn[standard]
```

Bold: Indicates a new term, an important word, or words that you see onscreen. For instance, words in menus or dialog boxes appear in **bold**. Here is an example: "Select **System info** from the **Administration** panel."

> **Tips or Important Notes**
> Appear like this.

Get in touch

Feedback from our readers is always welcome.

General feedback: If you have questions about any aspect of this book, email us at customercare@ packtpub.com and mention the book title in the subject of your message.

Errata: Although we have taken every care to ensure the accuracy of our content, mistakes do happen. If you have found a mistake in this book, we would be grateful if you would report this to us. Please visit www.packtpub.com/support/errata and fill in the form.

Piracy: If you come across any illegal copies of our works in any form on the internet, we would be grateful if you would provide us with the location address or website name. Please contact us at copyright@packt.com with a link to the material.

If you are interested in becoming an author: If there is a topic that you have expertise in and you are interested in either writing or contributing to a book, please visit authors.packtpub.com.

Share your thoughts

Once you've read *Building Python Microservices with FastAPI*, we'd love to hear your thoughts! Scan the QR code below to go straight to the Amazon review page for this book and share your feedback.

https://packt.link/r/1803245964

Your review is important to us and the tech community and will help us make sure we're delivering excellent quality content.

Part 1: Application-Related Architectural Concepts for FastAPI microservice development

In this part, we will look at the whole FastAPI framework and explore the systematic and ideal way of decomposing a monolithic application into several business units. During the process, you will see how to get started with development and what components there are in FastAPI that can be utilized to pursue microservices implementation.

This part comprises the following chapters:

- *Chapter 1, Setting Up FastAPI for Starters*
- *Chapter 2, Exploring the Core Components*
- *Chapter 3, Investigating Dependency Injection*
- *Chapter 4, Building the Microservice Application*

1
Setting Up FastAPI for Starters

In any software development work, it is always important to first know the business requirement of the project and the appropriate framework, tools, and deployment platform to use before pursuing the task. Frameworks that are easy to understand and use, seamless during coding, and within standards are always picked because of the integrity they provide to solve problems without risking too much development time. And a promising Python framework called **FastAPI**, created by *Sebastian Ramirez*, provides experienced developers, experts, and enthusiasts the best option for building **REST APIs** and microservices.

But before proceeding to the core details of building microservices using FastAPI, it is best to first learn the building blocks of this framework, such as how it captures clients' requests, how it builds the rules for each HTTP method, and how it manages HTTP responses. Learning the basic components is always essential to know the strengths and weaknesses of the framework and to what extent we can apply FastAPI to solve different enterprise-grade and microservices-related problems.

Thus, in this chapter, we're going to have a walkthrough of the basic features of FastAPI by covering the following main topics:

- The setup of the development environment
- Initialization and configuration of FastAPI
- Design and implementation of the REST APIs
- Managing user requests and server response
- Handling form parameters
- Handling cookies

Technical requirements

The software specimen for this chapter is a prototypical administrator-managed *online academic discussion forum*, which is an academic discussion hub where alumni, teachers, and students can exchange ideas. The prototype is working but it is open for changes, so you can tweak the code while reading this chapter. It is not designed to use any database management system, but all the data is temporarily stored in various Python collections. All the applications in this book are compiled and run using *Python 3.8*. Codes are all uploaded at `https://github.com/PacktPublishing/Building-Python-Microservices-with-FastAPI/tree/main/ch01`.

Setting up the development environment

The FastAPI framework is a fast, seamless, and robust Python framework but can only work on Python versions *3.6* and above. The **Integrated Development Environment** (**IDE**) used in this reference is **Visual Studio Code** (**VS Code**), which is an open source tool that we can download from this site: `https://code.visualstudio.com/`. Just be sure to install the VSC extensions such as *Python*, *Python for VS Code*, *Python Extension Pack*, *Python Indent*, and *Material Icon Theme* to provide your editor syntax checking, syntax highlighting, and other editor support.

After the successful installation of Python and VS Code, we can now install FastAPI using a terminal console. To ensure correct installation, first update Python's package installer (`pip`) by running this command:

```
python -m pip install --upgrade pip
```

Afterward, we install the framework by running this series of commands:

```
pip install fastapi
pip install uvicorn[standard]
pip install python-multipart
```

> **Important note**
>
> If you need to install the complete FastAPI platform, including all optional dependencies, the appropriate command is `pip install fastapi[all]`. Likewise, if you want to install and utilize the full-blown `uvicorn` server, you should run the `pip install uvicorn` command. Also, install the `bcrypt` module for encryption-related tasks.

At this point, you should have installed all the needed FastAPI module dependencies from the **pydantic** and **starlette** module components in your Python environment. Furthermore, the **python-multipart** module is required to create a REST API that handles form parameters. The installed `uvicorn`, however, is an ASGI-based server that will run your FastAPI applications. The **Asynchronous Server Gateway Interface** (**ASGI**) server that FastAPI uses makes it the fastest Python framework at the time of writing. The `uvicorn` server has the capability to run both synchronous and asynchronous services.

After the installation and configuration of the essential tools, modules, and IDE, let us now start our first API implementation using the framework.

Initializing and configuring FastAPI

Learning how to create applications using FastAPI is easy and straightforward. A simple application can be created just by creating a `main.py` file inside your `/ch01` project folder. In our *online academic discussion forum*, for instance, the application started with this code:

```
from fastapi import FastAPI
app = FastAPI()
```

This initializes the FastAPI framework. The application needs to instantiate the core `FastAPI` class from the `fastapi` module and use `app` as the reference variable to the object. Then, this object is used later as a Python `@app` decorator, which provides our application with some features such as *routes*, *middleware*, *exception handlers*, and *path operations*.

> **Important note**
> You can replace `app` with your preferred but valid Python variable name, such as `main_app`, `forum`, or `myapp`.

Now, your application is ready to manage REST APIs that are technically Python functions. But to declare them as REST service methods, we need to decorate them with the appropriate HTTP request method provided by the path operation `@app` decorator. This decorator contains the `get()`, `post()`, `delete()`, `put()`, `head()`, `patch()`, `trace()`, and `options()` path operations, which correspond to the eight HTTP request methods. And these *path operations* are decorated or annotated on top of the Python functions that we want to handle the request and response.

In our specimen, the first sample that the REST API created was this:

```
@app.get("/ch01/index")
def index():
    return {"message": "Welcome FastAPI Nerds"}
```

The preceding is a GET API service method that returns a JSON object. To locally run our application, we need to execute the following command:

```
uvicorn main:app --reload
```

This command will load the forum application to the uvicorn live server through the application's `main.py` file with FastAPI object referencing. Live reload is allowed by adding the `--reload` option, which enables the restart of the development server whenever there are changes in the code.

```
PS C:\Alibata\Training\Source\fastapi\ch01> uvicorn main:app --reload
INFO:      Uvicorn running on http://127.0.0.1:8000 (Press CTRL+C to quit)
INFO:      Started reloader process [2552] using watchgod
INFO:      Started server process [17880]
INFO:      Waiting for application startup.
INFO:      Application startup complete.
```

Figure 1.1 – The uvicorn console log

Figure 1.1 shows that uvicorn uses `localhost` to run the application with the default port `8000`. We can access our index page through `http://localhost:8000/ch01/index`. To stop the server, you just need to press the *Ctrl + C* keyboard keys.

After running our first endpoint, let us now explore how to implement the other types of HTTP methods, namely POST, DELETE, PUT, and PATCH.

Designing and implementing REST APIs

The **Representation State Transfer (REST)** API makes up the rules, processes, and tools that allow interaction among microservices. These are method services that are identified and executed through their endpoint URLs. Nowadays, focusing on API methods before building a whole application is one of the most popular and effective microservices design strategies. This approach, called an **API-first** microservices development, focuses first on the client's needs and then later identifies what API service methods we need to implement for these client requirements.

In our *online academic discussion forum* app, software functionality such as *user sign-up, login, profile management, message posting,* and *managing post replies* are some of the crucial needs we prioritized. In a FastAPI framework, these features are implemented as services using functions that are defined using Python's `def` keyword, with the association of the appropriate HTTP request method through the *path operations* provided by @app.

The `login` service, which requires `username` and `password` request parameters from the user, is implemented as a GET API method:

```
@app.get("/ch01/login/")
def login(username: str, password: str):
```

```
    if valid_users.get(username) == None:
        return {"message": "user does not exist"}
    else:
        user = valid_users.get(username)
        if checkpw(password.encode(),
                    user.passphrase.encode()):
            return user
        else:
            return {"message": "invalid user"}
```

This *login* service uses bcrypt's `checkpw()` function to check whether the password of the user is valid. Conversely, the *sign-up* service, which also requires user credentials from the client in the form of request parameters, is created as a POST API method:

```
@app.post("/ch01/login/signup")
def signup(uname: str, passwd: str):
    if (uname == None and passwd == None):
        return {"message": "invalid user"}
    elif not valid_users.get(uname) == None:
        return {"message": "user exists"}
    else:
        user = User(username=uname, password=passwd)
        pending_users[uname] = user
        return user
```

Among the *profile management* services, the following `update_profile()` service serves as a PUT API service, which requires the user to use an entirely new model object for profile information replacement and the client's username to serve as the key:

```
@app.put("/ch01/account/profile/update/{username}")
def update_profile(username: str, id: UUID,
                        new_profile: UserProfile):
    if valid_users.get(username) == None:
        return {"message": "user does not exist"}
    else:
        user = valid_users.get(username)
        if user.id == id:
            valid_profiles[username] = new_profile
```

```
                    return {"message": "successfully updated"}
            else:
                    return {"message": "user does not exist"}
```

Not all services that carry out updates are PUT API methods, such as the following update_profile_name() service, which only requires the user to submit a new first name, last name, and middle initial for partial replacement of a client's profile. This HTTP request, which is handier and more lightweight than a full-blown PUT method, only requires a PATCH action:

```
@app.patch("/ch01/account/profile/update/names/{username}")
def update_profile_names(username: str, id: UUID,
                                new_names: Dict[str, str]):
    if valid_users.get(username) == None:
        return {"message": "user does not exist"}
    elif new_names == None:
        return {"message": "new names are required"}
    else:
        user = valid_users.get(username)
        if user.id == id:
            profile = valid_profiles[username]
            profile.firstname = new_names['fname']
            profile.lastname = new_names['lname']
            profile.middle_initial = new_names['mi']
            valid_profiles[username] = profile
            return {"message": "successfully updated"}
        else:
            return {"message": "user does not exist"}
```

The last essential HTTP services that we included before building the application are the DELETE API methods. We use these services to delete records or information given a unique identification, such as username and a hashed id. An example is the following delete_post_discussion() service that allows a user to delete a posted discussion when given a username and the UUID (Universally Unique Identifier) of the posted message:

```
@app.delete("/ch01/discussion/posts/remove/{username}")
def delete_discussion(username: str, id: UUID):
    if valid_users.get(username) == None:
        return {"message": "user does not exist"}
    elif discussion_posts.get(id) == None:
```

```
        return {"message": "post does not exist"}
else:
        del discussion_posts[id]
        return {"message": "main post deleted"}
```

All path operations require a unique endpoint URL in the `str` format. A good practice is to start all URLs with the same top-level base path, such as `/ch01`, and then differ when reaching their respective subdirectories. After running the uvicorn server, we can check and validate whether all our URLs are valid and running by accessing the documentation URL, `http://localhost:8000/docs`. This path will show us a **OpenAPI** dashboard, as shown in *Figure 1.2*, listing all the API methods created for the application. Discussions on the OpenAPI will be covered in *Chapter 9, Utilizing Other Advanced Features*.

Figure 1.2 – A Swagger OpenAPI dashboard

After creating the endpoint services, let us scrutinize how FastAPI manages its incoming request body and the outgoing response.

Managing user requests and server response

Clients can pass their request data to FastAPI endpoint URLs through path parameters, query parameters, or headers to pursue service transactions. There are standards and ways to use these parameters to obtain incoming requests. Depending on the goal of the services, we use these parameters to influence and build the necessary responses the clients need. But before we discuss these various parameter types, let us explore first how we use *type hinting* in FastAPI's local parameter declaration.

Parameter type declaration

All request parameters are required to be type-declared in the method signature of the service method applying the **PEP 484** standard called *type hints*. FastAPI supports common types such as None, bool, int, and float and container types such as list, tuple, dict, set, frozenset, and deque. Other complex Python types such as datetime.date, datetime.time, datetime.datetime, datetime.delta, UUID, bytes, and Decimal are also supported.

The framework also supports the data types included in Python's typing module, responsible for *type hints*. These data types are standard notations for Python and variable type annotations that can help to pursue type checking and model validation during compilation, such as Optional, List, Dict, Set, Union, Tuple, FrozenSet, Iterable, and Deque.

Path parameters

FastAPI allows you to obtain request data from the endpoint URL of an API through a path parameter or path variable that makes the URL somewhat dynamic. This parameter holds a value that becomes part of a URL indicated by curly braces ({ }). After setting off these path parameters within the URL, FastAPI requires these parameters to be declared by applying *type hints*.

The following delete_user() service is a DELETE API method that uses a username path parameter to search for a login record for deletion:

```
@app.delete("/ch01/login/remove/{username}")
def delete_user(username: str):
    if username == None:
    return {"message": "invalid user"}
else:
    del valid_users[username]
    return {"message": "deleted user"}
```

Multiple path parameters are acceptable if the leftmost variables are more likely to be filled with values than the rightmost variables. In other words, the importance of the leftmost path variables will make the process more relevant and correct than those on the right. This standard is applied to ensure that the endpoint URL will not look like other URLs, which might cause some conflicts and confusion. The following `login_with_token()` service follows this standard, since `username` is a primary key and is as strong as, or even stronger than, its next parameter, `password`. There is an assurance that the URL will always look unique every time the endpoint is accessed because `username` will always be required, as well as `password`:

```
@app.get("/ch01/login/{username}/{password}")
def login_with_token(username: str, password:str,
                    id: UUID):
    if valid_users.get(username) == None:
        return {"message": "user does not exist"}
    else:
        user = valid_users[username]
        if user.id == id and checkpw(password.encode(),
                user.passphrase):
            return user
        else:
            return {"message": "invalid user"}
```

Unlike other web frameworks, FastAPI is not friendly with endpoint URLs that belong to base paths or top-level domain paths with different subdirectories. This occurrence happens when we have dynamic URL patterns that look the same as the other fixed endpoint URLs when assigned a specific path variable. These fixed URLs are implemented sequentially after these dynamic URLs. An example of these are the following services:

```
@app.get("/ch01/login/{username}/{password}")
def login_with_token(username: str, password:str,
                    id: UUID):
    if valid_users.get(username) == None:
        return {"message": "user does not exist"}
    else:
        user = valid_users[username]
        if user.id == id and checkpw(password.encode(),
                user.passphrase.encode()):
            return user
        else:
```

```
                    return {"message": "invalid user"}


@app.get("/ch01/login/details/info")
def login_info():
         return {"message": "username and password are
                                needed"}
```

This will give us an *HTTP Status Code 422* (*Unprocessable Entity*) when accessing `http://localhost:8080/ch01/login/details/info`. There should be no problem accessing the URL, since the API service is almost a stub or trivial JSON data. What happened in this scenario is that the fixed path's `details` and `info` path directories were treated as `username` and `password` parameter values, respectively. Because of confusion, the built-in data validation of FastAPI will show us a JSON-formatted error message that says, `{"detail":[{"loc":["query","id"],"msg":"field required","type":"value_error.missing"}]}`. To fix this problem, all fixed paths should be declared first before the dynamic endpoint URLs with path parameters. Thus, the preceding `login_info()` service should be declared first before `login_with_token()`.

Query parameters

A query parameter is a *key–value* pair supplied after the end of an endpoint URL, indicated by a question mark (?). Just like the path parameter, this also holds the request data. An API service can manage a series of query parameters separated by an ampersand (&). Like in path parameters, all query parameters are also declared in the service method. The following *login* service is a perfect specimen that uses query parameters:

```
@app.get("/ch01/login/")
def login(username: str, password: str):
    if valid_users.get(username) == None:
        return {"message": "user does not exist"}
    else:
        user = valid_users.get(username)
        if checkpw(password.encode(),
                user.passphrase.encode()):
            return user
        else:
            return {"message": "invalid user"}
```

The `login` service method uses `username` and `password` as query parameters in the `str` types. Both are required parameters, and assigning them with `None` as parameter values will give a compiler error.

FastAPI supports query parameters that are complex types, such as `list` and `dict`. But these Python collection types cannot specify the type of objects to store unless we apply the *generic type hints* for Python collections. The following `delete_users()` and `update_profile_names()` APIs use generic type hints, `List` and `Dict`, in declaring query parameters that are container types with type checking and data validation:

```python
from typing import Optional, List, Dict


@app.delete("/ch01/login/remove/all")
def delete_users(usernames: List[str]):
    for user in usernames:
        del valid_users[user]
    return {"message": "deleted users"}


@app.patch("/ch01/account/profile/update/names/{username}")
def update_profile_names(username: str, id: UUID,
                            new_names: Dict[str, str]):
    if valid_users.get(username) == None:
        return {"message": "user does not exist"}
    elif new_names == None:
        return {"message": "new names are required"}
    else:
        user = valid_users.get(username)
        if user.id == id:
            profile = valid_profiles[username]
            profile.firstname = new_names['fname']
            profile.lastname = new_names['lname']
            profile.middle_initial = new_names['mi']
            valid_profiles[username] = profile
            return {"message": "successfully updated"}
        else:
            return {"message": "user does not exist"}
```

FastAPI also allows you to explicitly assign default values to service function parameters.

Default parameters

There are times that we need to specify default values to the query parameter(s) and path parameter(s) of some API services to avoid validation error messages such as `field required` and `value_error.missing`. Setting default values to parameters will allow the execution of an API method with or without supplying the parameter values. Depending on the requirement, assigned default values are usually 0 for numeric types, `False` for bool types, empty string for string types, an empty list (`[]`) for List types, and an empty dictionary (`{}`) for `Dict` types. The following `delete pending users()` and `change_password()` services show us how to apply default values to the query parameter(s) and path parameter(s):

```python
@app.delete("/ch01/delete/users/pending")
def delete_pending_users(accounts: List[str] = []):
    for user in accounts:
        del pending_users[user]
    return {"message": "deleted pending users"}


@app.get("/ch01/login/password/change")
def change_password(username: str, old_passw: str = '',
                        new_passw: str = ''):
    passwd_len = 8
    if valid_users.get(username) == None:
        return {"message": "user does not exist"}
    elif old_passw == '' or new_passw == '':
        characters = ascii_lowercase
        temporary_passwd =
            ''.join(random.choice(characters) for i in
                    range(passwd_len))
        user = valid_users.get(username)
        user.password = temporary_passwd
        user.passphrase =
                    hashpw(temporary_passwd.encode(),gensalt())
        return user
    else:
        user = valid_users.get(username)
        if user.password == old_passw:
            user.password = new_passw
            user.passphrase = hashpw(new_pass.
encode(),gensalt())
```

```
            return user
        else:
            return {"message": "invalid user"}
```

delete_pending_users() can be executed even without passing any accounts argument, since accounts will be always an empty List by default. Likewise, change_password() can still continue its process without passing any old_passwd and new_passw, since they are both always defaulted to empty str. hashpw() is a bcrypt utility function that generates a hashed passphrase from an autogenerated *salt*.

Optional parameters

If the *path* and/or *query parameter(s)* of a service is/are not necessarily needed to be supplied by the user, meaning the API transactions can proceed with or without their inclusion in the request transaction, then we set them as *optional*. To declare an optional parameter, we need to import the Optional type from the typing module and then use it to set the parameter. It should wrap the supposed data type of the parameter using brackets ([]) and can have *any default value* if needed. Assigning the Optional parameter to a None value indicates that its exclusion from the parameter passing is allowed by the service, but it will hold a None value. The following services depict the use of optional parameters:

```
from typing import Optional, List, Dict


@app.post("/ch01/login/username/unlock")
def unlock_username(id: Optional[UUID] = None):
    if id == None:
        return {"message": "token needed"}
    else:
        for key, val in valid_users.items():
            if val.id == id:
                return {"username": val.username}
        return {"message": "user does not exist"}


@app.post("/ch01/login/password/unlock")
def unlock_password(username: Optional[str] = None,
                    id: Optional[UUID] = None):
    if username == None:
        return {"message": "username is required"}
    elif valid_users.get(username) == None:
```

```
                return {"message": "user does not exist"}
        else:
            if id == None:
                return {"message": "token needed"}
            else:
                user = valid_users.get(username)
                if user.id == id:
                    return {"password": user.password}
                else:
                    return {"message": "invalid token"}
```

In the *online academic discussion forum* application, we have services such as the preceding `unlock_username()` and `unlock_password()` services that declare all their parameters as `optional`. Just do not forget to apply exception handling or defensive validation in your implementation when dealing with these kinds of parameters to avoid *HTTP Status 500* (*Internal Server Error*).

> **Important note**
>
> The FastAPI framework does not allow you to directly assign the None value to a parameter just to declare an *optional* parameter. Although this is allowed with the old Python behavior, this is no longer recommended in the current Python versions for the purpose of built-in type checking and model validation.

Mixing all types of parameters

If you are planning to implement an API service method that declares optional, required, and default query and path parameters altogether, you can pursue it because the framework supports it, but approach it with some caution due to some standards and rules:

```
@app.patch("/ch01/account/profile/update/names/{username}")
def update_profile_names(id: UUID, username: str = '' ,
        new_names: Optional[Dict[str, str]] = None):
    if valid_users.get(username) == None:
        return {"message": "user does not exist"}
    elif new_names == None:
        return {"message": "new names are required"}
    else:
        user = valid_users.get(username)
        if user.id == id:
```

```
            profile = valid_profiles[username]
            profile.firstname = new_names['fname']
            profile.lastname = new_names['lname']
            profile.middle_initial = new_names['mi']
            valid_profiles[username] = profile
            return {"message": "successfully updated"}
        else:
            return {"message": "user does not exist"}
```

The updated version of the preceding update_profile_names() service declares a username path parameter, a UUID id query parameter, and an optional Dict[str, str] type. With mixed parameter types, all required parameters should be declared first, followed by default parameters, and last in the parameter list should be the optional types. Disregarding this ordering rule will generate a *compiler error*.

Request body

A *request body* is a body of data in bytes transmitted from a client to a server through a POST, PUT, DELETE, or PATCH HTTP method operation. In FastAPI, a service must declare a model object to represent and capture this request body to be processed for further results.

To implement a model class for the *request body*, you should first import the BaseModel class from the pydantic module. Then, create a subclass of it to utilize all the properties and behavior needed by the path operation in capturing the request body. Here are some of the data models used by our application:

```python
from pydantic import BaseModel

class User(BaseModel):
    username: str
    password: str

class UserProfile(BaseModel):
    firstname: str
    lastname: str
    middle_initial: str
    age: Optional[int] = 0
    salary: Optional[int] = 0
    birthday: date
    user_type: UserType
```

The attributes of the model classes must be explicitly declared by applying *type hints* and utilizing the common and complex data types used in the parameter declaration. These attributes can also be set as required, default, and optional, just like in the parameters.

Moreover, the `pydantic` module allows the creation of nested models, even the deeply nested ones. A sample of these is shown here:

```
class ForumPost(BaseModel):
    id: UUID
    topic: Optional[str] = None
    message: str
    post_type: PostType
    date_posted: datetime
    username: str


class ForumDiscussion(BaseModel):
    id: UUID
    main_post: ForumPost
    replies: Optional[List[ForumPost]] = None
    author: UserProfile
```

As seen in the preceding code, we have a `ForumPost` model, which has a `PostType` model attribute, and `ForumDiscussion`, which has a `List` attribute of `ForumPost`, a `ForumPost` model attribute, and a `UserProfile` attribute. This kind of model blueprint is called a *nested model approach*.

After creating these model classes, you can now *inject* these objects into the services that are intended to capture the *request body* from the clients. The following services utilize our `User` and `UserProfile` model classes to manage the request body:

```
@app.post("/ch01/login/validate", response_model=ValidUser)
def approve_user(user: User):
    if not valid_users.get(user.username) == None:
        return ValidUser(id=None, username = None,
            password = None, passphrase = None)
    else:
        valid_user = ValidUser(id=uuid1(),
            username= user.username,
            password = user.password,
            passphrase = hashpw(user.password.encode(),
```

```
                        gensalt()))
        valid_users[user.username] = valid_user
        del pending_users[user.username]
        return valid_user


@app.put("/ch01/account/profile/update/{username}")
def update_profile(username: str, id: UUID,
                   new_profile: UserProfile):
    if valid_users.get(username) == None:
        return {"message": "user does not exist"}
    else:
        user = valid_users.get(username)
        if user.id == id:
            valid_profiles[username] = new_profile
            return {"message": "successfully updated"}
        else:
            return {"message": "user does not exist"}
```

Models can be declared *required*, with a *default* instance value, or *optional* in the service method, depending on the specification of the API. Missing or incorrect details such as invalid password or None values in the approve_user() service will emit the *Status Code 500 (Internal Server Error)*. How FastAPI handles exceptions will be part of *Chapter 2, Exploring the Core Features*, discussions.

> **Important note**
>
> There are two essential points we need to emphasize when dealing with BaseModel class types. First, the pydantic module has a built-in JSON encoder that converts the JSON-formatted request body to the BaseModel object. So, there is no need create a custom converter to map the request body to the BaseModel model. Second, to instantiate a BaseModel class, all its required attributes must be initialized immediately through the constructor's named parameters.

Request headers

In a request-response transaction, it is not only the parameters that are accessible by the REST API methods but also the information that describes the context of the client where the request originated. Some common request headers such as User-Agent, Host, Accept, Accept-Language, Accept-Encoding, Referer, and Connection usually appear with request parameters and values during request transactions.

To access a request header, import first the `Header` function from the `fastapi` module. Then, declare the variable that has the same name as the header in the method service as `str` types and initialize the variable by calling the `Header(None)` function. The `None` argument enables the `Header()` function to declare the variable optionally, which is a best practice. For hyphenated request header names, the hyphen (-) should be converted to an underscore (_); otherwise, the Python compiler will flag a syntax error message. It is the task of the `Header()` function to convert the underscore (_) to a hyphen (-) during request header processing.

Our online academic discussion forum application has a `verify_headers()` service that retrieves core request headers needed to verify a client's access to the application:

```python
from fastapi import Header


@app.get("/ch01/headers/verify")
def verify_headers(host: Optional[str] = Header(None),
                   accept: Optional[str] = Header(None),
                   accept_language:
                       Optional[str] = Header(None),
                   accept_encoding:
                       Optional[str] = Header(None),
                   user_agent:
                       Optional[str] = Header(None)):
    request_headers["Host"] = host
    request_headers["Accept"] = accept
    request_headers["Accept-Language"] = accept_language
    request_headers["Accept-Encoding"] = accept_encoding
    request_headers["User-Agent"] = user_agent
    return request_headers
```

> **Important note**
> Non-inclusion of the `Header()` function call in the declaration will let FastAPI treat the variables as *query parameters*. Be cautious also with the spelling of the local parameter names, since they *are* the request header names per se except for the underscore.

Response data

All API services in FastAPI should return JSON data, or it will be invalid and may return `None` by default. These responses can be formed using `dict`, `BaseModel`, or `JSONResponse` objects. Discussions on `JSONResponse` will be discussed in the succeeding chapters.

The `pydantic` module's built-in JSON converter will manage the conversion of these custom responses to a JSON object, so there is no need to create a custom JSON encoder:

```
@app.post("/ch01/discussion/posts/add/{username}")
def post_discussion(username: str, post: Post,
                    post_type: PostType):
    if valid_users.get(username) == None:
        return {"message": "user does not exist"}
    elif not (discussion_posts.get(id) == None):
        return {"message": "post already exists"}
    else:
        forum_post = ForumPost(id=uuid1(),
          topic=post.topic, message=post.message,
          post_type=post_type,
          date_posted=post.date_posted, username=username)
        user = valid_profiles[username]
        forum = ForumDiscussion(id=uuid1(),
         main_post=forum_post, author=user, replies=list())
        discussion_posts[forum.id] = forum
        return forum
```

The preceding `post_discussion()` service returns two different hardcoded `dict` objects, with `message` as the key and an instantiated `ForumDiscussion` model.

On the other hand, this framework allows us to specify the return type of a service method. The setting of the return type happens in the `response_model` attribute of any of the `@app` path operations. Unfortunately, the parameter only recognizes `BaseModel` class types:

```
@app.post("/ch01/login/validate", response_model=ValidUser)
def approve_user(user: User):

    if not valid_users.get(user.username) == None:
        return ValidUser(id=None, username = None,
                   password = None, passphrase = None)
    else:
        valid_user = ValidUser(id=uuid1(),
         username= user.username, password = user.password,
          passphrase = hashpw(user.password.encode(),
                 gensalt()))
```

```
        valid_users[user.username] = valid_user
        del pending_users[user.username]
        return valid_user
```

The preceding `approve_user()` service specifies the required return of the API method, which is `ValidUser`.

Now, let us explore how FastAPI handles form parameters.

Handling form parameters

When API methods are designed to handle web forms, the services involved are required to retrieve form parameters instead of the request body because this form data is normally encoded as an `application/x-www-form-urlencoded` media type. These form parameters are conventionally `string` types, but the `pydantic` module's JSON encoder can convert each parameter value to its respective valid type.

All the form parameter variables can be declared *required*, with *default* values, or *optional* using the same set of Python types we used previously. Then, the `fastapi` module has a `Form` function that needs to be imported to initialize these form parameter variables during their declaration. To set these form parameters as *required*, the `Form()` function must have the ellipses (...) argument, thus calling it as `Form(...)`:

```
from fastapi import FastAPI, Form

@app.post("/ch01/account/profile/add",
                    response_model=UserProfile)
def add_profile(uname: str,
                fname: str = Form(...),
                lname: str = Form(...),
                mid_init: str = Form(...),
                user_age: int = Form(...),
                sal: float = Form(...),
                bday: str = Form(...),
                utype: UserType = Form(...)):
    if valid_users.get(uname) == None:
        return UserProfile(firstname=None, lastname=None,
            middle_initial=None, age=None,
            birthday=None, salary=None, user_type=None)
    else:
```

```
        profile = UserProfile(firstname=fname,
            lastname=lname, middle_initial=mid_init,
            age=user_age, birthday=datetime.strptime(bday,
              '%m/%d/%Y'), salary=sal, user_type=utype)
    valid_profiles[uname] = profile
    return profile
```

The preceding `add_profile()` service shows us how to call the `Form(...)` function to return a Form object during the parameter declaration.

> **Important note**
>
> Form-handling services will not work if the `python-multipart` module is not installed.

Sometimes, we need browser cookies to establish an identity for our application, leave trails in the browser for every user transaction, or store product information for a purpose. If FastAPI can manage form data, it can also do the same with cookies.

Managing cookies

A *cookie* is a piece of information stored in the browser to pursue some purpose, such as login user authorization, web agent response generation, and session handling-related tasks. One cookie is always a key-value pair that are both string types.

FastAPI allows services to create cookies individually through the `Response` library class from its `fastapi` module. To use it, it needs to appear as the first local parameter of the service, but we do not let the application or client pass an argument to it. Using the dependency injection principle, the framework will provide the `Response` instance to the service and not the application. When the service has other parameters to declare, the additional declaration should happen right after the declaration of the `Response` parameter.

The `Response` object has a `set_cookie()` method that contains two required named parameters: the *key*, which sets the cookie name, and the *value*, which stores the cookie value. This method only generates one cookie and stores it in the browser afterward:

```
@app.post("/ch01/login/rememberme/create/")
def create_cookies(resp: Response, id: UUID,
                  username: str = ''):
    resp.set_cookie(key="userkey", value=username)
    resp.set_cookie(key="identity", value=str(id))
    return {"message": "remember-me tokens created"}
```

The preceding `create_cookies()` method shows us the creation of *remember-me tokens* such as `userkey` and `identity` for the *remember-me* authorization of our *online academic discussion forum* project.

To retrieve these cookies, local parameters that have the same name as the cookies are declared in the service method as `str` types, since cookie values are always strings. As with `Header` and `Form`, the `fastapi` module also provides a `Cookie` function that is needed to initialize each declared cookie parameter variable. The `Cookie()` function should always have the `None` argument to set the parameters optionally, ensuring that the API method executes without problems whenever the headers are not present in the request transaction. The following `access_cookie()` service retrieves all the *remember-me* authorization cookies created by the previous service:

```
@app.get("/ch01/login/cookies")
def access_cookie(userkey: Optional[str] = Cookie(None),
            identity: Optional[str] = Cookie(None)):
    cookies["userkey"] = userkey
    cookies["identity"] = identity
    return cookies
```

Summary

This chapter is essential to familiarize ourselves with FastAPI and understand its basic components. The concept that we can get from this chapter can measure how much adjustment and effort we need to invest into translating or rewriting some existing applications to FastAPI. Knowing its basics will help us learn how to install its modules, structure the project directories, and learn the core library classes and functions needed to build a simple enterprise-grade application.

With the help of our recipe *online academic discussion forum* application, this chapter showed us how to build different REST APIs associated with HTTP methods using the FastAPI module class and Python `def` functions. From there, we learned how to capture incoming request data and headers using the local parameters of the API methods and how these API methods should return a response to the client. And through this chapter, we saw how easy it is for FastAPI to capture form data from `<form></form>` of any UI templates and that is using the `Form` function. Aside from the `Form` function, the FastAPI module also has the `Cookie` function to help us create and retrieve cookies from the browser, and `Header` to retrieve the request header part of an incoming request transaction.

Overall, this chapter has prepared us for advanced discussions that will center on other features of FastAPI that can help us upgrade our simple applications to full-blown ones. The next chapter will cover these essential core features, which will provide our application with the needed response encoder and generator, exception handlers, middleware, and other components related to asynchronous transactions.

2

Exploring the Core Features

In the previous chapter, we found out how easy it is to install and start developing REST APIs using the **FastAPI** framework. Handling requests, cookies, and form data was fast, easy, and straightforward with FastAPI, as was building the different HTTP path operations.

To learn about the framework's features further, this chapter will guide us on how to upgrade our REST APIs by adding some essential FastAPI features to the implementation. These include some handlers that can help minimize unchecked exceptions, JSON encoders that can directly manage endpoint responses, background jobs that can create audit trails and logs, and multiple threads to run some API methods asynchronously with the **uvicorn**'s main thread. Moreover, issues such as managing source files, modules, and packages for huge enterprise projects will also be addressed in this chapter. This chapter will use and dissect an *intelligent tourist system* prototype to assist with elaborating upon and exemplifying FastAPI's core modules.

Based on these aforementioned features, this chapter will discuss the following major concepts that can help us extend our learning about this framework:

- Structuring and organizing huge projects
- Managing API-related exceptions
- Converting objects to JSON-compatible types
- Managing API responses
- Creating background processes
- Using asynchronous path operations
- Applying middleware to filter path operations

Technical requirements

This chapter will implement a prototype of an intelligent tourist system designed to provide booking information and reservation about tourist spots. It can provide user details, tourist spot details, and location grids. It also allows users or tourists to comment on tours and rate them. The prototype has an administrator account for adding and removing all the tour details, managing users, and providing some listings. The application will not use any database management system yet, so all the data is temporarily stored in Python collections. The code is all uploaded at `https://github.com/PacktPublishing/Building-Python-Microservices-with-FastAPI/tree/main/ch02`.

Structuring and organizing huge projects

In FastAPI, big projects are organized and structured by adding *packages* and *modules* without destroying the setup, configuration, and purpose. The project should always be flexible and scalable in case of additional features and requirements. One component must correspond to one package, with several modules equivalent to a *blueprint* in a Flask framework.

In this prototypical intelligent tourist system, the application has several modules such as the login, administration, visit, destination, and feedback-related functionalities. The two most crucial are the *visit* module, which manages all the travel bookings of the users, and the *feedback* module, which enables clients to post their feedback regarding their experiences at every destination. These modules should be separated from the rest since they provide the core transactions. *Figure 2.1* shows how to group implementations and separate a module from the rest using **packages**:

Figure 2.1 – The FastAPI project structure

Each package in *Figure 2.1* contains all the modules where the API services and some dependencies are implemented. All the aforementioned modules now have their own respective packages that make it easy to test, debug, and expand the application. Testing FastAPI components will be discussed in the upcoming chapters.

> **Important note**
>
> FastAPI does not require adding the __init__.py file into each Python package when using *VS Code Editor* and *Python 3.8* during development, unlike in Flask. The __pycache__ folder generated inside a package during compilation contains binaries of the module scripts accessed and utilized by other modules. The main folder will also become a package since it will have its own __pycache__ folder together with the others. But we must exclude __pycache__ when deploying the application to the repository, since it may take up a lot of space.

On the other hand, what remains in the main folder are the core components such as the *background tasks, custom exception handlers, middleware,* and the main.py file. Now, let us learn about how FastAPI can bundle all these packages as one huge application when deployed.

Implementing the API services

For these module packages to function, the main.py file must call and register all their API implementations through the FastAPI instance. The scripts inside each package are already REST API implementations of the microservices, except that they are built by APIRouter instead of the FastAPI object. APIRouter also has the same path operations, query and request parameter setup, handling of form data, generation of responses, and parameter injection of model objects. What is lacking in APIRouter is the support for an exception handler, middleware declaration, and customization:

```
from fastapi import APIRouter
from login.user import Signup, User, Tourist,
       pending_users, approved_users


router = APIRouter()


@router.get("/ch02/admin/tourists/list")
def list_all_tourists():
    return approved_users
```

The list_all_tourists() API method operation here is part of the manager.py module in the admin package, implemented using APIRouter due to project structuring. The method returns a list of tourist records that are allowed to access the application, which can only be provided by the user.py module in the login package.

Importing the module components

Module scripts can share their *containers*, `BaseModel` *classes*, and other *resource objects* to other modules using Python's `from... import` statement. Python's `from... import` statement is better since it allows us to import specific components from a module, instead of including unnecessary ones:

```python
from fastapi import APIRouter, status
from places.destination import Tour, TourBasicInfo,
    TourInput, TourLocation, tours, tours_basic_info,
    tours_locations


router = APIRouter()


@router.put("/ch02/admin/destination/update",
            status_code=status.HTTP_202_ACCEPTED)
def update_tour_destination(tour: Tour):
    try:
        tid = tour.id
        tours[tid] = tour
        tour_basic_info = TourBasicInfo(id=tid,
            name=tour.name, type=tour.type,
            amenities=tour.amenities, ratings=tour.ratings)
        tour_location = TourLocation(id=tid,
            name=tour.name, city=tour.city,
            country=tour.country, location=tour.location )
        tours_basic_info[tid] = tour_basic_info
        tours_locations[tid] = tour_location
        return { "message" : "tour updated" }
    except:
        return { "message" : "tour does not exist" }
```

The `update_tour_destination()` operation here will not work without importing the `Tour`, `TourBasicInfo`, and `TourLocation` model classes from `destination.py` in the `places` package. It shows the dependency between modules that happens when structuring is imposed on big enterprise web projects.

Module scripts can also import components from the main project folder when needed by the implementation. One such example is accessing the *middleware*, *exception handlers*, and *tasks* from the `main.py` file.

> **Important note**
>
> Avoid cycles when dealing with the `from... import` statement. A **cycle** happens when a module script, `a.py`, accesses components from `b.py` that import resource objects from `a.py`. FastAPI does not accept this scenario and will issue an error message.

Implementing the new main.py file

Technically, the project's packages and its module scripts will not be recognized by the framework unless their respective `router` object is added or injected into the application's core through the `main.py` file. `main.py`, just as the other project-level scripts do, uses `FastAPI` and not `APIRouter` to create and register components, as well as the package's modules. The FastAPI class has an `include_router()` method that adds all these routers and injects them into the framework to make them part of the project structure. Beyond registering the routers, this method can also add other attributes and components to the router such as *URL prefixes*, *tags*, *dependencies such as exception handlers*, and *status codes*:

```python
from fastapi import FastAPI, Request

from admin import manager
from login import user
from feedback import post
from places import destination
from tourist import visit

app = FastAPI()

app.include_router(manager.router)
app.include_router(user.router)
app.include_router(destination.router)
app.include_router(visit.router)
app.include_router(
    post.router,
    prefix="/ch02/post"
)
```

This code is the `main.py` implementation of the intelligent tourist system prototype tasked to import all the registers of the module's scripts from the different packages, before adding them as components to the framework. Run the application using the following command:

```
uvicorn main:app --reload
```

This will allow you to access all the APIs of these modules at `http://localhost:8000/docs`.

What happens to the application when API services encounter runtime problems during execution? Is there a way to manage these problems besides applying Python's `try-except` block? Let us explore implementing API services with exception-handling mechanisms further.

Managing API-related exceptions

The FastAPI framework has a built-in exception handler derived from its Starlette toolkit that always returns default JSON responses whenever `HTTPException` is encountered during the execution of the REST API operation. For instance, accessing the API at `http://localhost:8000/ch02/user/login` without providing the `username` and `password` will give us the default JSON output depicted in *Figure 2.2*:

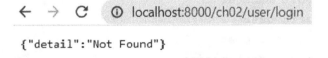

```
{"detail":"Not Found"}
```

Figure 2.2 – The default exception result

In some rare cases, the framework sometimes chooses to return the HTTP response status instead of the default JSON content. But developers can still opt to override these default handlers to choose which responses to return whenever a specific exception cause happens.

Let us now explore how to formulate a standardized and appropriate way of managing runtime errors in our API implementation.

A single status code response

One way of managing the exception-handling mechanism of your application is to apply a `try-except` block to manage the return responses of your API when it encounters an exception or none. After applying `try-block`, the operation should trigger a single **status code**, most often **Status Code 200 (SC 200)**. The path operation of `FastAPI` and `APIRouter` has a `status_code` parameter that we can use to indicate the type of status code we want to raise.

In FastAPI, status codes are integer constants that are found in the `status` module. It also allows integer literals to indicate the needed status code if they are a valid status code number.

> **Important Note**
>
> A status code is a 3-digit number that indicates a reason for, information on, or status of the HTTP response of a REST API operation. The status code range 200 to 299 denotes a successful response, 300 to 399 pertains to redirection, 400-499 pertains to client-related problems, and 500 to 599 is related to server errors.

This technique is rarely used because there are times that an operation needs to be clear in recognizing every exception that it encounters, which can only be done by returning HTTPException instead of a custom error message wrapped in a JSON object:

```python
from fastapi import APIRouter, status


@router.put("/ch02/admin/destination/update",
                status_code=status.HTTP_202_ACCEPTED)
def update_tour_destination(tour: Tour):
    try:
        tid = tour.id
        tours[tid] = tour
        tour_basic_info = TourBasicInfo(id=tid,
            name=tour.name, type=tour.type,
            amenities=tour.amenities, ratings=tour.ratings)
        tour_location = TourLocation(id=tid,
            name=tour.name, city=tour.city,
            country=tour.country, location=tour.location )
        tours_basic_info[tid] = tour_basic_info
        tours_locations[tid] = tour_location
        return { "message" : "tour updated" }
    except:
        return { "message" : "tour does not exist" }


@router.get("/ch02/admin/destination/list",
                status_code=200)
def list_all_tours():
    return tours
```

The `list_all_tours()` method shown here is the kind of REST API service that should emit Status Code 200 – it gives an error-free result just by rendering the Python collection with data. Observe that the literal integer value, 200, or *SC 200*, assigned to the `status_code` parameter of the GET path operation always raises an *OK* status. On the other hand, the `update_tour_destination()` method shows another approach in emitting status codes by using a `try-except` block, wherein both blocks return a custom JSON response. Whichever scenario happens, it will always trigger *SC 202*, which may not apply to some REST implementations. After the `status` module is imported, its `HTTP_202_ACCEPTED` constant is used to set the value of the `status_code` parameter.

Multiple status codes

If we need each block in `try-except` to return their respective status code, we need to avoid using the `status_code` parameter of the path operations and use `JSONResponse` instead. `JSONResponse` is one of the FastAPI classes used to render a JSON response to the client. It is instantiated, constructor-injected with values for its `content` and `status_code` parameters, and returned by the path operations. By default, the framework uses this API to help path operations render responses as JSON types. Its `content` parameter should be a JSON-type object, while the `status_code` parameter can be an integer constant and a valid status code number, or it can be a constant from the module status:

```
from fastapi.responses import JSONResponse


@router.post("/ch02/admin/destination/add")
add_tour_destination(input: TourInput):
    try:
        tid = uuid1()
        tour = Tour(id=tid, name=input.name,
            city=input.city, country=input.country,
            type=input.type, location=input.location,
            amenities=input.amenities, feedbacks=list(),
            ratings=0.0, visits=0, isBooked=False)
        tour_basic_info = TourBasicInfo(id=tid,
            name=input.name, type=input.type,
            amenities=input.amenities, ratings=0.0)
        tour_location = TourLocation(id=tid,
            name=input.name, city=input.city,
            country=input.country, location=input.location )
        tours[tid] = tour
        tours_basic_info[tid] = tour_basic_info
```

```
            tours_locations[tid] = tour_location
            tour_json = jsonable_encoder(tour)
            return JSONResponse(content=tour_json,
                status_code=status.HTTP_201_CREATED)
        except:
            return JSONResponse(
              content={"message" : "invalid tour"},
              status_code=status.HTTP_500_INTERNAL_SERVER_ERROR)
```

The add_tour_destination() operation here has a try-except block where its try block returns the tour details and *SC 201*, while its catch block returns an error message inside a JSON-type object with a server error of *SC 500*.

Raising HTTPException

Another way of managing possible errors is by letting the REST API throw the HTTPException object. HTTPException is a FastAPI class that has required constructor parameters: detail, which needs an error message in the str type, and status_code, which asks for a valid integer value. The detail part is converted to JSON-type and returned to the user as a response after the HTTPException instance is thrown by the operation.

To throw HTTPException, a validation process using any variations of if statements is more appropriate than using the try-except block because the cause of the error needs to be identified before throwing the HTTPException object using the raise statement. Once raise is executed, the whole operation will halt and send the HTTP error message in JSON-type to the client with the specified status code:

```
from fastapi import APIRouter, HTTPException, status


@router.post("/ch02/tourist/tour/booking/add")
def create_booking(tour: TourBasicInfo, touristId: UUID):
    if approved_users.get(touristId) == None:
        raise HTTPException(status_code=500,
            detail="details are missing")
    booking = Booking(id=uuid1(), destination=tour,
      booking_date=datetime.now(), tourist_id=touristId)
    approved_users[touristId].tours.append(tour)
    approved_users[touristId].booked += 1
    tours[tour.id].isBooked = True
```

```
        tours[tour.id].visits += 1
        return booking
```

The `create_booking()` operation here simulates a booking process for a *tourist* account, but before the procedure starts, it first checks whether the *tourist* is still a valid user; otherwise, it will raise `HTTPException`, halting all the operations in order to return an error message.

Custom exceptions

It is also possible to create a user-defined `HTTPException` object to handle business-specific problems. This custom exception requires a custom handler needed to manage its response to the client whenever an operation raises it. These custom components should be available to all API methods across the project structure; thus, they must be implemented at the project-folder level.

In our application, there are two custom exceptions created in `handler_exceptions.py`, the `PostFeedbackException` and `PostRatingFeedback` exceptions, which handle problems related to posting feedback and ratings on a particular tour:

```python
from fastapi import FastAPI, Request, status, HTTPException


class PostFeedbackException(HTTPException):
    def __init__(self, detail: str, status_code: int):
        self.status_code = status_code
        self.detail = detail


class PostRatingException(HTTPException):
    def __init__(self, detail: str, status_code: int):
        self.status_code = status_code
        self.detail = detail
```

A valid FastAPI exception is a subclass of an `HTTPException` object inheriting the essential attributes, namely the `status_code` and `detail` attributes. We need to supply values to these attributes before the path operation raises the exception. After creating these custom exceptions, a specific handler is implemented and mapped to an exception.

The FastAPI @app decorator in `main.py` has an `exception_handler()` method, used to define a custom handler and map it to the appropriate custom exception. A handler is simply a Python function with two local parameters, `Request` and the *custom exception* that it manages. The purpose of the `Request` object is to retrieve cookies, payloads, headers, query parameters, and path parameters from the path operation if the handler expects any of this request data. Now, once the custom exception is raised, the handler is set to generate a JSON-type response to the client containing the `detail` and the `status_code` attributes provided by the path operation that raised the exception:

```
from fastapi.responses import JSONResponse
from fastapi import FastAPI, Request, status, HTTPException


@app.exception_handler(PostFeedbackException)
def feedback_exception_handler(req: Request,
        ex: PostFeedbackException):
    return JSONResponse(
        status_code=ex.status_code,
        content={"message": f"error: {ex.detail}"}
        )


@app.exception_handler(PostRatingException)
def rating_exception_handler(req: Request,
        ex: PostRatingException):
    return JSONResponse(
        status_code=ex.status_code,
        content={"message": f"error: {ex.detail}"}
        )
```

When an operation in `post.py` raises `PostFeedbackException`, the `feedback_exception_handler()` given here will trigger its execution to generate a response that can provide details about what has caused the feedback problem. The same thing will happen to `PostRatingException` and its `rating_exception_handler()`:

```
from handlers import PostRatingException,
                        PostFeedbackException


@router.post("/feedback/add")
def post_tourist_feedback(touristId: UUID, tid: UUID,
```

```
        post: Post, bg_task: BackgroundTasks):
    if approved_users.get(touristId) == None and
         tours.get(tid) == None:
        raise PostFeedbackException(detail='tourist and
              tour details invalid', status_code=403)
    assessId = uuid1()
    assessment = Assessment(id=assessId, post=post,
         tour_id= tid, tourist_id=touristId)
    feedback_tour[assessId] = assessment
    tours[tid].ratings = (tours[tid].ratings +
                          post.rating)/2
    bg_task.add_task(log_post_transaction,
         str(touristId), message="post_tourist_feedback")
    assess_json = jsonable_encoder(assessment)
    return JSONResponse(content=assess_json,
                        status_code=200)


@router.post("/feedback/update/rating")
def update_tour_rating(assessId: UUID,
            new_rating: StarRating):
    if feedback_tour.get(assessId) == None:
        raise PostRatingException(
         detail='tour assessment invalid', status_code=403)
    tid = feedback_tour[assessId].tour_id
    tours[tid].ratings = (tours[tid].ratings +
                          new_rating)/2
    tour_json = jsonable_encoder(tours[tid])
    return JSONResponse(content=tour_json, status_code=200)
```

post_tourist_feedback() and update_tour_rating() here are the API operations that will raise the PostFeedbackException and PostRatingException custom exceptions, respectively, triggering the execution of their handlers. The detail and status_code values injected into the constructor are passed to the handlers to create the response.

A default handler override

The optimum way to override the exception-handling mechanism of your application is to replace the global exception handler of the FastAPI framework that manages its core Starlette's `HTTPException` and the `RequestValidationError` triggered by **Pydantic**'s request validation process. For instance, if we want to change the response format of the global exception sent to the client using `raise` from JSON-type to plain text, we can create custom handlers for each of the aforementioned core exceptions that will pursue the format conversion. The following snippets of `main.py` show these types of custom handlers:

```
from fastapi.responses import PlainTextResponse
from starlette.exceptions import HTTPException as
        GlobalStarletteHTTPException
from fastapi.exceptions import RequestValidationError
from handler_exceptions import PostFeedbackException,
        PostRatingException


@app.exception_handler(GlobalStarletteHTTPException)
def global_exception_handler(req: Request,
                ex: str
    return PlainTextResponse(f"Error message:
        {ex}", status_code=ex.status_code)


@app.exception_handler(RequestValidationError)
def validationerror_exception_handler(req: Request,
                ex: str
    return PlainTextResponse(f"Error message:
        {str(ex)}", status_code=400)
```

Both the `global_exception_handler()` and `validationerror_exception_handler()` handlers are implemented to change the framework's JSON-type exception response to `PlainTextResponse`. An alias, `GlobalStarletteHTTPException`, is assigned to Starlette's `HTTPException` class to distinguish it from FastAPI's `HTTPException`, which we previously used to build custom exceptions. On the other hand, `PostFeedbackException` and `PostRatingException` are both implemented in the `handler_exceptions.py` module.

JSON objects are all over the FastAPI framework's REST API implementation, from the incoming request to the outgoing responses. However, what if the JSON data involved in the process is not a FastAPI JSON-compatible type? The following discussion will expound more upon this kind of object.

Converting objects to JSON-compatible types

It is easier for FastAPI to process JSON-compatible types such as `dict`, `list`, and `BaseModel` objects because they can be easily converted to JSON by the framework using its default JSON editor. However, there are circumstances in which runtime exceptions are raised when processing BaseModel, data model, or JSON objects containing data. One of the many reasons for this is that these data objects have attributes that are not supported by JSON rules, such as UUID and non-built-in date types. Regardless, using a framework's module classes, these objects can still be utilized by converting them into JSON-compatible ones.

When it comes to the direct handling of the API operation's responses, FastAPI has a built-in method that can encode typical model objects to convert them to JSON-compatible types before persisting them to any datastore or passing them to the `detail` parameter of `JSONResponse`. This method, `jsonable_encoder()`, returns a `dict` type with all the keys and values compatible with JSON:

```python
from fastapi.encoders import jsonable_encoder
from fastapi.responses import JSONResponse


class Tourist(BaseModel):
    id: UUID
    login: User
    date_signed: datetime
    booked: int
    tours: List[TourBasicInfo]


@router.post("/ch02/user/signup/")
async def signup(signup: Signup):
    try:
        userid = uuid1()
        login = User(id=userid, username=signup.username,
                password=signup.password)
        tourist = Tourist(id=userid, login=login,
            date_signed=datetime.now(), booked=0,
            tours=list() )
        tourist_json = jsonable_encoder(tourist)
        pending_users[userid] = tourist_json
        return JSONResponse(content=tourist_json,
            status_code=status.HTTP_201_CREATED)
    except:
```

```
    return JSONResponse(content={"message":
    "invalid operation"},
    status_code=status.HTTP_500_INTERNAL_SERVER_ERROR)
```

Our application has a POST operation, `signup()`, shown here that captures the profile of a newly created user to be approved by the administrator. If you observe the `Tourist` model class, it has a `date_signed` attribute that is declared as `datettime`, and temporal types are not always JSON-friendly. Having model objects with non-JSON-friendly components in FastAPI-related operations can cause serious exceptions. To avoid these Pydantic validation issues, it is always advisable to use `jsonable_encoder()` to manage the conversion of all the attributes of our model object into JSON-types.

> **Important note**
>
> The `json` module with its `dumps()` and `loads()` utility methods can be used instead of `jsonable_encoder()` but a custom JSON encoder should be created to successfully map the `UUID` type, the formatted `date` type, and other complex attribute types to `str`.

Chapter 9, Utilizing Other Advanced Features, will discuss other JSON encoders that can encode and decode JSON responses faster than the `json` module.

Managing API responses

The use of `jsonable_encoder()` can help an API method not only with data persistency problems but also with the integrity and correctness of its response. In the `signup()` service method, `JSONResponse` returns the encoded `Tourist` model instead of the original object to ensure that the client always received a JSON response. Aside from raising status codes and providing error messages, `JSONResponse` can also do some tricks in handling the API responses to the client. Although optional in many circumstances, applying the encoder method when generating responses is recommended to avoid runtime errors:

```
from fastapi.encoders import jsonable_encoder
from fastapi.responses import JSONResponse


@router.get("/ch02/destinations/details/{id}")
def check_tour_profile(id: UUID):
    tour_info_json = jsonable_encoder(tours[id])
    return JSONResponse(content=tour_info_json)
```

`check_tour_profile()` here uses `JSONResponse` to ensure that its response is JSON-compatible and is fetched from the purpose of managing its exceptions. Moreover, it can also be used to return headers together with the JSON-type response:

```
@router.get("/ch02/destinations/list/all")
def list_tour_destinations():
    tours_json = jsonable_encoder(tours)
    resp_headers = {'X-Access-Tours': 'Try Us',
        'X-Contact-Details':'1-900-888-TOLL',
        'Set-Cookie':'AppName=ITS; Max-Age=3600; Version=1'}
    return JSONResponse(content=tours_json,
        headers=resp_headers)
```

The application's `list_tour_destinations()` here returns three cookies: AppName, Max-Age, and Version, and two user-defined response headers. Headers that have names beginning with X- are custom headers. Besides `JSONResponse`, the `fastapi` module also has a `Response` class that can create response headers:

```
from fastapi import APIRouter, Response


@router.get("/ch02/destinations/mostbooked")
def check_recommended_tour(resp: Response):
    resp.headers['X-Access-Tours'] = 'TryUs'
    resp.headers['X-Contact-Details'] = '1900888TOLL'
    resp.headers['Content-Language'] = 'en-US'
    ranked_desc_rates = sort_orders = sorted(tours.items(),
        key=lambda x: x[1].ratings, reverse=True)
    return ranked_desc_rates;
```

Our prototype's `check_recommend_tour()` uses `Response` to create two custom response headers and a known **Content-Language**. Always remember that headers are all `str` types and are stored in the browser for many reasons, such as creating an identity for the application, leaving user trails, dropping advertisement-related data, or leaving an error message to the browser when an API encounters one:

```
@router.get("/ch02/tourist/tour/booked")
def show_booked_tours(touristId: UUID):
    if approved_users.get(touristId) == None:
        raise HTTPException(
        status_code=status.HTTP_500_INTERNAL_SERVER_ERROR,
```

```
            detail="details are missing",
        headers={"X-InputError":"missing tourist ID"})
    return approved_users[touristId].tours
```

HTTPException, as shown in the show_booked_tours() service method here, not only contains the status code and error message but also some headers in case the operation needs to leave some error information to the browser once it is raised.

Let us now explore the capability of FastAPI to create and manage transactions that are designed to run in the background using some server threads.

Creating background processes

The FastAPI framework is also capable of running background jobs as part of an API service execution. It can even run more than one job almost simultaneously without intervening in the main service execution. The class responsible for this is BackgroundTasks, which is part of the fastapi module. Conventionally, we declare this at the end of the parameter list of the API service method for the framework to inject the BackgroundTask instance.

In our application, the task is to create audit logs of all API service executions and store them in an audit_log.txt file. This operation is part of the background.py script that is part of the main project folder, and the code is shown here:

```
from datetime import datetime


def audit_log_transaction(touristId: str, message=""):
    with open("audit_log.txt", mode="a") as logfile:
        content = f"tourist {touristId} executed {message}
            at {datetime.now()}"
        logfile.write(content)
```

Here, audit_log_transaction() must be injected into the application using BackgroundTasks's add_task() method to become a background process that will be executed by the framework later:

```
from fastapi import APIRouter, status, BackgroundTasks


@router.post("/ch02/user/login/")
async def login(login: User, bg_task:BackgroundTasks):
    try:
        signup_json =
            jsonable_encoder(approved_users[login.id])
```

```python
        bg_task.add_task(audit_log_transaction,
            touristId=str(login.id), message="login")
        return JSONResponse(content=signup_json,
            status_code=status.HTTP_200_OK)
    except:
        return JSONResponse(
        content={"message": "invalid operation"},
        status_code=status.HTTP_500_INTERNAL_SERVER_ERROR)


@router.get("/ch02/user/login/{username}/{password}")
async def login(username:str, password: str,
                    bg_task:BackgroundTasks):
    tourist_list = [ tourist for tourist in
        approved_users.values()
            if tourist['login']['username'] == username and
                tourist['login']['password'] == password]
    if len(tourist_list) == 0 or tourist_list == None:
        return JSONResponse(
            content={"message": "invalid operation"},
            status_code=status.HTTP_403_FORBIDDEN)
    else:
        tourist = tourist_list[0]
        tour_json = jsonable_encoder(tourist)
        bg_task.add_task(audit_log_transaction,
            touristId=str(tourist['login']['id']),
message="login")
        return JSONResponse(content=tour_json,
            status_code=status.HTTP_200_OK)
```

The login() service method is just one of the services of our application that logs its details. It uses the bg_task object to add audit_log_transaction() into the framework to be processed later. Transactions such as logging, *SMTP-/FTP*-related requirements, events, and some database-related triggers are the best candidates for background jobs.

> **Important Note**
>
> Clients will always get their response from the REST API method despite the execution time of the background task. Background tasks are for processes that will take enough time that including them in the API operation could cause performance degradation.

Using asynchronous path operations

When it comes to improving performance, FastAPI is an asynchronous framework, and it uses Python's **AsyncIO** principles and concepts to create a REST API implementation that can run separately and independently from the application's main thread. The idea also applies to how a background task is executed. Now, to create an asynchronous REST endpoint, attach `async` to the `func` signature of the service:

```
@router.get("/feedback/list")
async def show_tourist_post(touristId: UUID):
    tourist_posts = [assess for assess in feedback_tour.
values()
             if assess.tourist_id == touristId]
    tourist_posts_json = jsonable_encoder(tourist_posts)
    return JSONResponse(content=tourist_posts_json,
                    status_code=200)
```

Our application has a `show_tourist_post()` service that can retrieve all the feedback posted by a certain `touristId` about a vacation tour that they have experienced. The application will not be affected no matter how long the service will take because its execution will be simultaneous to the `main` thread.

> **Important Note**
>
> The `feedback` APIRouter uses a `/ch02/post` prefix indicated in its `main.py`'s `include_router()` registration. So, to run `show_tourist_post()`, the URL should be `http://localhost:8000/ch02/post`.

An asynchronous API endpoint can invoke both synchronous and asynchronous Python functions that can be DAO (Data Access Object), native services, or utility. Since FastAPI also follows the `Async/Await` design pattern, the asynchronous endpoint can call an asynchronous non-API operation using the `await` keyword, which halts the API operation until the non-API transaction is done processing a promise:

```
from utility import check_post_owner


@router.delete("/feedback/delete")
async def delete_tourist_feedback(assessId: UUID,
             touristId: UUID ):
    if approved_users.get(touristId) == None and
           feedback_tour.get(assessId):
        raise PostFeedbackException(detail='tourist and
```

```
                    tour details invalid', status_code=403)      post_
    delete = [access for access in feedback_tour.values()
                  if access.id == assessId]
        for key in post_delete:
            is_owner = await check_post_owner(feedback_tour,
                          access.id, touristId)
            if is_owner:
                del feedback_tour[access.id]
        return JSONResponse(content={"message" : f"deleted
            posts of {touristId}"}, status_code=200)
```

`delete_tourist_feedback()` here is an asynchronous REST API endpoint that calls an asynchronous Python function, `check_post_owner()`, from the `utility.py` script. For the two components to have a handshake, the API service invokes `check_post_owner()`, using an `await` keyword for the former to wait for the latter to finish its validation, and retrieves the promise that it can get from `await`.

> **Important Note**
>
> The `await` keyword can only be used with the `async` REST API and native transactions, not with synchronous ones.

To improve performance, you can add more threads within the `uvicorn` thread pool by including the `--workers` option when running the server. Indicate your preferred number of threads after calling the option:

```
uvicorn main:app --workers 5 --reload
```

Chapter 8, Creating Coroutines, Events, and Message-Driven Transactions, will discuss the *AsyncIO* platform and the use of *coroutines* in more detail.

And now, the last, most important core feature that FastAPI can provide is the middleware or the "request-response filter."

Applying middleware to filter path operations

There are FastAPI components that are inherently asynchronous and one of them is the middleware. It is an asynchronous function that acts as a filter for the REST API services. It filters out the incoming request to pursue validation, authentication, logging, background processing, or content generation from the cookies, headers, request parameters, query parameters, form data, or authentication details of the request body before it reaches the API service method. Equally, it takes the outgoing response body to pursue rendition change, response header updates and additions, and other kinds of transformation that could possibly be applied to the response before it reaches the client. Middleware should be implemented at the project level and can even be part of main.py:

```
@app.middleware("http")
async def log_transaction_filter(request: Request,
             call_next):
    start_time = datetime.now()
    method_name= request.method
    qp_map = request.query_parasms
    pp_map = request.path_params
    with open("request_log.txt", mode="a") as reqfile:
        content = f"method: {method_name}, query param:
            {qp_map}, path params: {pp_map} received at
            {datetime.now()}"
        reqfile.write(content)
    response = await call_next(request)
    process_time = datetime.now() - start_time
    response.headers["X-Time-Elapsed"] = str(process_time)
    return response
```

To implement middleware, first, create an async function that has two local parameters: the first one is Request and the second one is a function called call_next(), which takes the Request parameter as its argument to return the response. Then, decorate the method with @app.middleware("http") to inject the component into the framework.

The tourist application has one middleware implemented by the asynchronous add_transaction_filter() here that logs the necessary request data of a particular API method before its execution and modifies its response object by adding a response header, X-Time-Elapsed, which carries the running time of the execution.

The execution of await call_next(request) is the most crucial part of the middleware because it explicitly controls the execution of the REST API service. It is the area of the component where Request passes through to the API execution for processing. Equally, it is where Response tunnels out, going to the client.

Besides logging, middleware can also be used for implementing one-way or two-way authentication, checking user roles and permissions, global exception handling, and other filtering-related operations right before the execution of call_next(). When it comes to controlling the outgoing Response, it can be used to modify the content type of the response, remove some existing browser cookies, modify the response detail and status code, redirections, and other response transformation-related transactions. *Chapter 9, Utilizing Other Advanced Features*, will discuss the types of middleware, middleware chaining, and other means to customize middleware to help build a better microservice.

> **Important note**
> The FastAPI framework has some built-in middleware that is ready to be injected into the application such as GzipMiddleware, ServerErrorMiddleware, TrustedHostMiddleware, ExceptionMiddleware, CORSMiddleware, SessionMiddleware, and HTTPSRedirectionMiddleware.

Summary

Exploring the core details of a framework always helps us create a comprehensive plan and design to build quality applications to the required standards. We have learned that FastAPI injects all its incoming form data, request parameters, query parameters, cookies, request headers, and authentication details into the Request object, and the outgoing cookies, response headers, and response data are carried out to the client by the Response object. When managing the response data, the framework has a built-in jsonable_encoder() function that can convert the model into JSON types to be rendered by the JSONResponse object. Its middleware is one powerful feature of FastAPI because we can customize it to handle the Request object before it reaches the API execution and the Response object before the client receives it.

Managing the exceptions is always the first step to consider before creating a practical and sustainable solution for the resiliency and health of a microservice architecture. Alongside its robust default **Starlette** global exception handler and **Pydantic** model validator, FastAPI allows exception-handling customization that provides the flexibility needed when business processes become intricate.

FastAPI follows Python's **AsyncIO** principles and standards for creating async REST endpoints, which makes implementation easy, handy, and reliable. This kind of platform is helpful for building complex architectures that require more threads and asynchronous transactions.

This chapter is a great leap toward fully learning about the principles and standards of how FastAPI manages its web containers. The features highlighted in this chapter hitherto open up a new level of knowledge that we need to explore further if we want to utilize FastAPI to build great microservices. In the next chapter, we will be discussing FastAPI dependency injection and how this design pattern affects our FastAPI projects.

3

Investigating Dependency Injection

Since the first chapter, **Dependency Injection (DI)** has been accountable for building clean and robust FastAPI REST services and was apparent in some of our sample APIs that required `BaseModel`, `Request`, `Response`, and `BackgroundTasks` in their processes. Applying DI proves that instantiating some FastAPI classes is not always the ideal approach, since the framework has a built-in container that can provide the objects of these classes for the API services. This method of object management makes FastAPI easy and efficient to use.

FastAPI has a container where the DI policy is applied to instantiate module classes and even functions. We only need to specify and declare these module APIs to the services, middleware, authenticator, data sources, and test cases because the rest of the object assembly, management, and instantiation is now the responsibility of the built-in container.

This chapter will help you to understand how to manage objects needed by the application, such as minimizing some instances and creating loose bindings among them. Knowing the effectiveness of DI on FastAPI is the first step in designing our microservice applications. Our discussions will focus on the following:

- Applying **Inversion of Control (IoC)** and DI
- Exploring ways of injecting dependencies
- Organizing a project based on dependencies
- Using third-party containers
- Scoping of instances

Technical requirements

This chapter uses a software prototype called *online recipe system*, which manages, evaluates, rates, and reports recipes of different types and origins. Applying a DI pattern is the priority of this project, so expect some changes in the development strategies and approaches, such as adding model, repository, and service folders. This software is for food enthusiasts or chefs who want to share their specialties, newbies looking for recipes to experiment with, and guests who just like to browse through different food menus. This open-ended application does not use any database management system yet, so all the data is temporarily stored in Python containers. Code is all uploaded at https://github.com/PacktPublishing/Building-Python-Microservices-with-FastAPI/tree/main/ch03.

Applying IoC/DI

FastAPI is a framework that supports the IoC principle, which means that it has a container that can instantiate objects for an application. In a typical programming scenario, we instantiate classes to use them in many ways to build a running application. But with IoC, the framework instantiates the components for the application. *Figure 3.1* shows the whole picture of the IoC principle and the participation of one of its forms, called the DI.

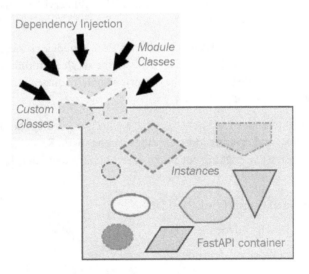

Figure 3.1 – The IoC principle

For FastAPI, DI is not only a principle but a mechanism to integrate an object into a component that leads to creating a *loosely coupled* but *highly cohesive* software structure. Almost all components can be candidates for DI, including *functions*. But for now, let us focus on *callable components* that provide some JSON objects once they are injected into an API service – injectable and callable components that we call *dependency functions*.

Injecting a dependency function

A **dependency function** is a typical Python function that has parameters such as a path operation or an API service and a return value of a JSON type too. A sample implementation is `create_login()`, as shown in the following code, which is in the project's `/api/users.py` module:

```
def create_login(id:UUID, username: str, password:str,
                   type: UserType):
    account = {"id": id, "username": username, "password":
                   password, "type": type}
    return account
```

The function requires the `id`, `username`, and `password` parameters and `type` to continue its process and return a valid JSON `account` object, derived from these parameters. A dependency function sometimes uses some underlying formula, resource, or complex algorithms to derive its function value, but for now, we utilize it as a *placeholder of data* or `dict`.

Common to dependency functions are method parameters that serve as placeholders to a REST API's incoming request. These are wired into the API's method parameter list as a domain model to the query parameters or request body through DI. The `Depends()` function from the `fastapi` module pursues the injection before it wires the injectable to a local parameter. The module function can only take one parameter for injection:

```
from fastapi import APIRouter, Depends

@router.get("/users/function/add")
def populate_user_accounts(
                   user_account=Depends(create_login)):
    account_dict = jsonable_encoder(user_account)
    login = Login(**account_dict)
    login_details[login.id] = login
    return login
```

The preceding is a code fragment from our *online recipe system* that shows how `Depends()` injects `create_login()` into the framework's container and fetches its instance for *wiring* to the `populate_user_accounts()` service. Syntactically, the injection process only needs the name of the function dependency without the parenthesis. Again, the purpose of `create_login()` is to capture the query parameters of the API service. The `jsonable_encoder()` is very useful to many APIs in converting these injectables to JSON-compatible types such as `dict`, which are essential for *instantiating the needed data models* and *generating responses*.

> **Important Note**
>
> The term **dependencies** can be used interchangeably with **injectables**, **dependables**, **resources**, **providers**, or **components**.

Injecting a callable class

FastAPI also allows classes to be injected into any components, since they can also be considered *callable* components. A class becomes *callable* during instantiation when the call to its constructor, __init__(self), is done. Some of these classes have *no-arg* constructors, while others, such as the following Login class, require constructor arguments:

```
class Login:
    def __init__(self, id: UUID, username: str,
                 password: str, type: UserType):
        self.id = id
        self.username = username
        self.password = password
        self.type= type
```

The Login class, located in /model/users.py, needs id, username, password, and type passed to its constructor before the instantiation. A possible instantiation would be Login(id=' 249a0837-c52e-48cd-bc19-c78e6099f931', username='admin', password='admin2255', type=UserType.admin). Overall, we can observe the similarity between a class and a dependency function based on their callable behavior and the ability to capture request data, such as a model attribute.

Conversely, the populate_login_without_service() shown in the following code block shows how Depends() injects Login to the service. The Depends() function tells the built-in container to instantiate Login and fetches that instance, ready to be assigned to the user_account local parameter:

```
@router.post("/users/datamodel/add")
def populate_login_without_service(
             user_account=Depends(Login)):
    account_dict = jsonable_encoder(user_account)
    login = Login(**account_dict)
    login_details[login.id] = login
    return login
```

> **Important Note**
>
> All dependencies should be declared at the right-most part of the service's parameter list. If there are query, path, or form parameters, injectables should come last. Moreover, the use of the `jsonable_encoder()` function can be an option if the *injectables* do not contain data that are hard to encode by default.

Building nested dependencies

There are some scenarios when injectables are also dependent on other dependencies. When we inject a function dependency to another function, a class dependency to another class injectable, or a function resource to a class, the goal is to build nested dependencies. Nested dependencies are beneficial to REST APIs, with lengthy and complex request data that needs structuring and grouping through sub-domain models. These sub-domains or domain models within a model are later encoded as sub-dependencies into JSON-compatible types by FastAPI:

```python
async def create_user_details(id: UUID, firstname: str,
        lastname: str, middle: str, bday: date, pos: str,
        login=Depends(create_login)):
    user = {"id": id, "firstname": firstname,
            "lastname": lastname, "middle": middle,
            "bday": bday, "pos": pos, "login": login}
    return user
```

The preceding asynchronous `create_user_details()` function shows that even a dependency function needs another dependency to satisfy its purpose. This function is dependent on `create_login()`, which is another dependable component. With this nested dependency setup, wiring the `create_user_details()` into an API service also includes the injection of `create_login()` into the container. In short, there is a chain of DIs that will be created when nested dependencies are applied:

```python
@router.post("/users/add/profile")
async def add_profile_login(
        profile=Depends(create_user_details)):
    user_profile = jsonable_encoder(profile)
    user = User(**user_profile)
    login = user.login
    login = Login(**login)
    user_profiles[user.id] = user
```

```
        login_details[login.id] = login
        return user_profile
```

The preceding `add_profile_login()` service provides a clear picture of its dependency on `create_user_details()`, including its underlying login details. The FastAPI container successfully created the two functions through chained DI to capture the request data during the API's request transactions.

Conversely, a class can also be dependable on another class. An example is the `Profile` class, shown here, which is dependent on `UserDetails` and `Login` classes:

```python
class Login:
    def __init__(self, id: UUID, username: str,
                    password: str, type: UserType):
        self.id = id
        self.username = username
        self.password = password
        self.type= type


class UserDetails:
    def __init__(self, id: UUID, firstname: str,
            lastname: str, middle: str, bday: date,
                pos: str ):
        self.id = id
        self.firstname = firstname
        self.lastname = lastname
        self.middle = middle
        self.bday = bday
        self.pos = pos


class Profile:
    def __init__(self, id: UUID, date_created: date,
        login=Depends(Login), user=Depends(UserDetails)):
        self.id = id
        self.date_created = date_created
        self.login = login
        self.user = user
```

There is a nested dependency here because two classes will be injected altogether once `Profile` is wired to a REST API service.

A clear advantage of these chained dependencies is depicted in the following `add_profile_login_models()` service:

```
@router.post("/users/add/model/profile")
async def add_profile_login_models(
                profile=Depends(Profile)):
    user_details = jsonable_encoder(profile.user)
    login_details = jsonable_encoder(profile.login)
    user = UserDetails(**user_details)
    login = Login(**login_details)
    user_profiles[user.id] = user
    login_details[login.id] = login
    return {"profile_created": profile.date_created}
```

The extraction of `profile.user` and `profile.login` makes it easier for the service to identify what query data to deserialize. It also helps the service determine which group of data needs `Login` instantiation and which is for `UserDetails`. As a result, it will be easier to manage the persistency of these objects in their respective `dict` repositories.

Creating explicit dependencies between function and class will be discussed later, but for now, let us examine how to fine-tune whenever we use lots of these nested dependencies in our application.

Caching the dependencies

All dependencies are *cacheable*, and FastAPI caches all these dependencies during a request transaction. If a dependable is common to all services, FastAPI will not allow you to fetch these objects from its container *by default*. Rather, it will look for this injectable from its cache to be used multiple times across the API layer. Saving dependencies, especially nested ones, is a good feature of FastAPI because it optimizes the performance of the REST service.

Conversely, `Depends()` has a `use_cache` parameter that we can set to `False` if we want to bypass this caching mechanism. Configuring this hook will not save the dependencies from the cache during request transactions, allowing `Depends()` to fetch the instances from the container more frequently. Another version of the `add_profile_login_models()` service, shown here, shows how to disable dependency caching:

```
@router.post("/users/add/model/profile")
async def add_profile_login_models(
```

```
profile:Profile=Depends(Profile, use_cache=False)):
    user_details = jsonable_encoder(profile.user)
    login_details = jsonable_encoder(profile.login)
    ... ... ... ... ... ...
    return {"profile_created": profile.date_created}
```

Another obvious change in the preceding service implementation is the presence of the `Profile` data type in the local parameter declaration. Is this really allowed by FastAPI?

Declaring Depends() parameter types

Generally, we do not declare types of the local parameters that will reference the injected dependencies. Due to *type hints*, we can optionally associate the references with their appropriate object types. For instance, we can re-implement `populate_user_accounts()` to include the type of `user_account`, such as the following one:

```
@router.get("/users/function/add")
def populate_user_accounts(
        user_account:Login=Depends(create_login)):
    account_dict = jsonable_encoder(user_account)
    login = Login(**account_dict)
    login_details[login.id] = login
    return login
```

This scenario happens very rarely, since `create_login()` is a dependency function, and we usually do not create classes only to provide the blueprint type of its returned values. But when we use class dependables, declaring the appropriate class type to the wired object is feasible, as in the following `add_profile_login_models()` service, which declares the `profile` parameter as `Profile`:

```
@router.post("/users/add/model/profile")
async def add_profile_login_models(
        profile:Profile=Depends(Profile)):
    user_details = jsonable_encoder(profile.user)
    login_details = jsonable_encoder(profile.login)
    ... ... ... ... ... ...
    return {"profile_created": profile.date_created}
```

Although the declaration is syntactically valid, the expression looks *repetitive* and *redundant* because the `Profile` type appears twice in the declaration portion. To avoid this redundancy, we can replace the statement with a shorthand version by omitting the class name inside the `Depends()` function. Thus, a better way of declaring the preceding `profile` should be the following:

```
@router.post("/users/add/model/profile")
async def add_profile_login_models(
                profile:Profile=Depends()):
    user_details = jsonable_encoder(profile.user)
    ... ... ... ... ... ...
    return {"profile_created": profile.date_created}
```

The changes reflected on the parameter list will not affect the performance of the request transaction of the `add_profile_login_models()` service.

Injecting asynchronous dependencies

A FastAPI built-in container does not only manage *synchronous* function dependables but also *asynchronous* ones. The following `create_user_details()` is an asynchronous dependency, read to be wired to a service:

```
async def create_user_details(id: UUID, firstname: str,
        lastname: str, middle: str, bday: date, pos: str,
        login=Depends(create_login)):
    user = {"id": id, "firstname": firstname,
            "lastname": lastname, "middle": middle,
            "bday": bday, "pos": pos, "login": login}
    return user
```

The container can manage both synchronous and asynchronous function dependency. It can allow the wiring of *asynchronous dependables* on an asynchronous API service or some *asynchronous ones* on a synchronous API. In cases where the dependency and the services are both asynchronous, applying the `async/await` protocol is recommended to avoid discrepancies in the results. `create_user_details()`, which is dependent on a synchronous `create_login()`, is wired on `add_profile_login()`, which is an asynchronous API.

After learning how DI design pattern works in FastAPI, the next step is to know the different levels of strategy for applying `Depends()` in our application.

Exploring ways of injecting dependencies

From the previous discussions, we know that FastAPI has a built-in container through which some objects are injected and instantiated. Likewise, we have learned that the only FastAPI components that are injectables are those so-called dependables, injectables, or dependencies. Now, let us enumerate different ways to pursue the DI pattern in our application.

Dependency injection on services

The most common area where DI occurs is in the *parameter list* of a service method. Any discussions regarding this strategy have been tackled already in the previous examples, so we only need to present additional points concerning this strategy:

- First, the number of custom injectables a service method should take is also part of the concern. When it comes to complex query parameters or request bodies, API services can take more than one injectable as long as there are no similar instance variable names among these dependables. This *variable name collision* among the dependables will lead to having one parameter entry for the conflicted variable during the request transactions, thus sharing the same value for all these conflicting variables.

- Second, the appropriate *HTTP method operation* to work with the injectables is also one aspect to consider. Both function and class dependencies can work with the GET, POST, PUT, and PATCH operations, except for those dependables with attribute types such as numeric Enum and UUID that can cause an *HTTP Status 422* (**Unprocessable Entity**) due to conversion problems. We must plan which HTTP method is applicable for some dependable(s) first, before implementing the service method.

- Third, not all dependables are placeholders of request data. Unlike the class dependables, dependency functions are not specifically used for returning objects or `dict`. Some of them are used in *filtering request data*, *scrutinizing authentication details*, *managing form data*, *verifying header values*, *handling cookies*, and *throwing some errors* when there are violations of some rules. The following `get_all_recipes()` service is dependent on a `get_recipe_service()` injectable that will query all the recipes from the `dict` repository of the application:

```
@router.get("/recipes/list/all")
def get_all_recipes(handler=Depends(get_recipe_service)):
    return handler.get_recipes()
```

The dependency function provides the needed transactions such as saving and retrieving records of recipes. Instead of the usual instantiation or method call, a better strategy is to inject these *dependable services* into the API implementation. The `handler` method parameter, which refers to the instance of `get_recipe_service()`, invokes the `get_recipes()` transactions of a particular service to retrieve all the menus and ingredients stored in the repository.

Dependency injection on path operators

There is always an option to implement *triggers*, *validators*, and *exception handlers* as injectable functions. Since these dependables work like *filters* to the incoming request, their injection happens in the *path operator* and not in the service parameter list. The following code is an implementation of the `check_feedback_length()` validator, found in `/dependencies/posts.py`, which checks whether the feedback posted by a user regarding a recipe should be at least 20 characters, including spaces:

```
def check_feedback_length(request: Request):
    feedback = request.query_params["feedback"]
    if feedback == None:
        raise HTTPException(status_code=500,
            detail="feedback does not exist")
    if len(feedback) < 20:
        raise HTTPException(status_code=403,
            detail="length of feedback … not lower … 20")
```

The validator pauses the API execution to retrieve the feedback from a post to be validated *if its length is lower than 20*. If the dependency function finds it `True`, it will throw an *HTTP Status 403*. Alternatively, it will emit a *Status Code 500* if the feedback is missing from the request data; otherwise, it will let the API transaction finish its task.

Compared to the `create_post()` and `post_service()` dependables, the following script shows that the `check_feedback_length()` validator is not invoked anywhere inside the `insert_post_feedback()` service:

```
async def create_post(id:UUID, feedback: str,
    rating: RecipeRating, userId: UUID, date_posted: date):
    post = {"id": id, "feedback": feedback,
            "rating": rating, "userId" : userId,
            "date_posted": date_posted}
    return post
```

```
@router.post("/posts/insert",
        dependencies=[Depends(check_feedback_length)])
async def insert_post_feedback(post=Depends(create_post),
            handler=Depends(post_service)):
    post_dict = jsonable_encoder(post)
    post_obj = Post(**post_dict)
    handler.add_post(post_obj)
    return post
```

The validator will always work closely with the incoming request transaction, whereas the other two injectables, post and handler, are part of the API's transactions.

> **Important Note**
>
> The path router of APIRouter can accommodate more than one injectable, which is why its dependencies parameter always needs a List value ([]).

Dependency injection on routers

However, some transactions are not localized to work on one specific API. There are dependency functions created to work with a certain group of REST API services within an application, such as the following count_user_by_type() and handler check_credential_error() events that are designed to manage incoming requests of REST APIs under the user.router group. This strategy requires DI at the APIRouter level:

```
from fastapi import Request, HTTPException
from repository.aggregates import stats_user_type
import json

def count_user_by_type(request: Request):
    try:
        count =
            stats_user_type[request.query_params.get("type")]
        count += 1
        stats_user_type[request.query_params.get("type")] =
            count
        print(json.dumps(stats_user_type))
    except:
```

```
        stats_user_type[request.query_params.get("type")] = 1

def check_credential_error(request: Request):
    try:
        username = request.query_params.get("username")
        password = request.query_params.get("password")
        if username == password:
          raise HTTPException(status_code=403,
            detail="username should not be equal to password")
    except:
        raise HTTPException(status_code=500,
            detail="encountered internal problems")
```

Based on the preceding implementations, the goal of count_user_by_type() is to build an updated frequency of users in stats_user_type according to UserType. Its execution starts right after the REST API receives a new user and login details from the client. While it checks the UserType of the new record, the API service pauses briefly and resumes after the function dependency has completed its tasks.

Conversely, the task of check_credential_error() is to ensure that the username and password of the new user should not be the same. It throws an *HTTP Status 403* when the credentials are the same, which will halt the whole REST service transaction.

Injecting these two dependables through APIRouter means that all the REST API services registered in that APIRouter will always trigger the executions of these dependencies. The dependencies can only work with API services designed to persist the like of the user and login details, as shown here:

```
from fastapi import APIRouter, Depends
router = APIRouter(dependencies=[
                    Depends(count_user_by_type),
                    Depends(check_credential_error)])


@router.get("/users/function/add")
def populate_user_accounts(
        user_account:Login=Depends(create_login)):
    account_dict = jsonable_encoder(user_account)
    login = Login(**account_dict)
    login_details[login.id] = login
    return login
```

check_credential_error(), which is injected into the APIRouter component, filters the username and password derived from the create_login() injectable function. Likewise, it filters the create_user_details() injectable of the add_profile_login() service, as shown in the following snippet:

```
@router.post("/users/add/profile")
async def add_profile_login(
        profile=Depends(create_user_details)):
    user_profile = jsonable_encoder(profile)
    user = User(**user_profile)
    login = user.login
    login = Login(**login)
    user_profiles[user.id] = user
    login_details[login.id] = login
    return user_profile


@router.post("/users/datamodel/add")
def populate_login_without_service(
        user_account=Depends(Login)):
    account_dict = jsonable_encoder(user_account)
    login = Login(**account_dict)
    login_details[login.id] = login
    return login
```

The Login injectable class also undergoes filtering through check_credential_error(). It also contains the username and password parameters that the injectable function can filter. Conversely, the injectable Profile of the following add_profile_login_models() service is not excluded from the error-checking mechanism because it has a Login dependency in its constructor. Having the Login dependable means check_cedential_error() will also filter Profile.

With check_credential_error() is the count_user_by_type() injectable that counts the number of users that access the API service:

```
@router.post("/users/add/model/profile")
async def add_profile_login_models(
        profile:Profile=Depends(Profile)):
    user_details = jsonable_encoder(profile.user)
    ... ... ... ... ... ...
```

```
login = Login(**login_details)
user_profiles[user.id] = user
login_details[login.id] = login
return {"profile_created": profile.date_created}
```

A dependency function wired into `APIRouter` should apply defensive programming and a proper `try-except` to avoid parameter conflicts with the API services. If we were to run `check_credential_error()` with a `list_all_user()` service, for instance, expect some runtime problems because there is no `login` persistence involved during data retrieval.

> **Important Note**
> Like its path operators, the constructor of `APIRouter` can also accept more than one injectable because its `dependencies` parameter will allow a `List ([])` of valid ones.

Dependency injection on main.py

There are portions of the software that are very hard to automate because of their vast and complex scopes, so considering them will always be a waste of time and effort. These *cross-cutting concerns* span from the UI level down to the data tier, which explains why these functionalities are impractical and even inconceivable to manage and implement using typical programming paradigms. These cross-cutting concerns are transactions such as *exception logging*, *caching*, *instrumentation*, and *user authorization*, common to any application.

FastAPI has an easy remedy to address these *features*: to create them as injectables to the FastAPI instance of `main.py`:

```
from fastapi import Request
from uuid import uuid1


service_paths_log = dict()


def log_transaction(request: Request):
    service_paths_log[uuid1()] = request.url.path
```

The preceding `log_transaction()` is a simple logger of the URL paths invoked or accessed by the client. While the application is running, this cross-cut should propagate the repository with different URLs coming from any `APIRouter`. This task can only happen when we inject this function through the FastAPI instance of `main.py`:

```
from fastapi import FastAPI, Depends
from api import recipes, users, posts, login, admin,
```

```
        keywords, admin_mcontainer, complaints
from dependencies.global_transactions import
        log_transaction


app = FastAPI(dependencies=[Depends(log_transaction)])


app.include_router(recipes.router, prefix="/ch03")
app.include_router(users.router, prefix="/ch03")
    ... ... ... ... ... ...
app.include_router(admin.router, prefix="/ch03")
app.include_router(keywords.router, prefix="/ch03")
app.include_router(admin_mcontainer.router, prefix="/ch03")
app.include_router(complaints.router, prefix="/ch03")
```

Dependencies *auto-wired* to the FastAPI constructor are known as *global dependencies* because they are accessible by any REST APIs from the routers. For instance, log_transaction(), depicted in the preceding script, will execute every time the APIs from the recipes, users, posts, or complaints routers process their respective request transactions.

> **Important Note**
> Like APIRouter, the constructor of FastAPI allows more than function dependency.

Aside from these strategies, DI can also help us organize our application by having repository, service, and model layers.

Organizing a project based on dependencies

It is feasible to use a *repository-service* pattern in some complex FastAPI applications through DI. The repository-service pattern is responsible for the creation of the **repository layer** of the application, which manages the Creation, Reading, Updates, and Deletion (CRUD) of data source. A repository layer requires **data models** that depict the table schemas of a collection or database. The repository layer needs the **service layer** to establish a connection with other parts of the application. The service layer operates like a business layer, where the data sources and business processes meet to derive all the necessary objects needed by the REST API. The communication between the repository and service layers can only be possible by creating injectables. Now, let us explore how the layers shown in *Figure 3.2* are built by DI using injectable components.

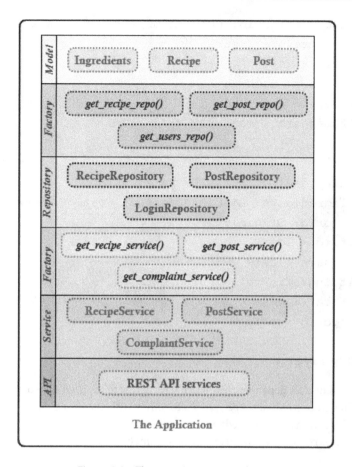

Figure 3.2 – The repository-service layers

The model layer

This layer is purely composed of *resources*, *collections*, and *Python classes* that can be used by the repository layer to create CRUD transactions. Some model classes are dependable on other models, but some are just independent blueprints designed for data placeholder. Some of the application's model classes that store recipe-related details are shown here:

```
from uuid import UUID
from model.classifications import Category, Origin
from typing import Optional, List


class Ingredient:
```

```
    def __init__(self, id: UUID, name:str, qty : float,
             measure : str):
        self.id = id
        self.name = name
        self.qty = qty
        self.measure = measure


class Recipe:
    def __init__(self, id: UUID, name: str,
             ingredients: List[Ingredient], cat: Category,
             orig: Origin):
        self.id = id
        self.name = name
        self.ingredients = ingredients
        self.cat = cat
        self.orig = orig
```

The repository layer

This layer is composed of class dependencies, which have access to the *data store* or improvised `dict` repositories, just like in our *online recipe system*. Together with the model layer, these repository classes build the CRUD transactions needed by the REST API. The following is an implementation of `RecipeRepository` that has two transactions, namely `insert_recipe()` and `query_recipes()`:

```
from model.recipes import Recipe
from model.recipes import Ingredient
from model.classifications import Category, Origin
from uuid import uuid1


recipes = dict()


class RecipeRepository:
    def __init__(self):
        ingrA1 = Ingredient(measure='cup', qty=1,
             name='grape tomatoes', id=uuid1())
        ingrA2 = Ingredient(measure='teaspoon', qty=0.5,
             name='salt', id=uuid1())
```

```
ingrA3 = Ingredient(measure='pepper', qty=0.25,
     name='pepper', id=uuid1())
... ... ... ... ... ...
recipeA = Recipe(orig=Origin.european ,
 ingredients= [ingrA1, ingrA2, ingrA3, ingrA4,
     ingrA5, ingrA6, ingrA7, ingrA8, ingrA9],
 cat= Category.breakfast,
 name='Crustless quiche bites with asparagus and
     oven-dried tomatoes',
 id=uuid1())


ingrB1 = Ingredient(measure='tablespoon', qty=1,
   name='oil', id=uuid1())
ingrB2 = Ingredient(measure='cup', qty=0.5,
   name='chopped tomatoes', id=uuid1())
... ... ... ... ... ...


recipeB = Recipe(orig=Origin.carribean ,
   ingredients= [ingrB1, ingrB2, ingrB3, ingrB4,
     ingrB5],
   cat= Category.breakfast,
   name='Fried eggs, Caribbean style', id=uuid1())


ingrC1 = Ingredient(measure='pounds', qty=2.25,
   name='sweet yellow onions', id=uuid1())
ingrC2 = Ingredient(measure='cloves', qty=10,
   name='garlic', id=uuid1())
... ... ... ... ... ...
recipeC = Recipe(orig=Origin.mediterranean ,
   ingredients= [ingrC1, ingrC2, ingrC3, ingrC4,
     ingrC5, ingrC6, ingrC7, ingrC8],
   cat= Category.soup,
   name='Creamy roasted onion soup', id=uuid1())

recipes[recipeA.id] = recipeA
recipes[recipeB.id] = recipeB
```

```
        recipes[recipeC.id] = recipeC

    def insert_recipe(self, recipe: Recipe):
        recipes[recipe.id] = recipe

    def query_recipes(self):
        return recipes
```

Its constructor is used to populate the recipes with some initial data. The constructor of an *injectable repository* class plays a role in datastore setup and configuration, and this is where we *auto-wire* dependency if there is any. Conversely, the implementation includes two Enum classes – `Category` and `Origin` – which provide lookup values to the recipe's menu category and place of origin respectively.

The repository factory methods

This layer uses the *factory design pattern* to add a more loose coupling design between the repository and service layer. Although this approach is optional, this is still an option to manage the threshold of interdependency between the two layers, especially when there are frequent changes in the performance, processes, and results of the CRUD transactions. The following are the repository factory methods used by our application:

```
def get_recipe_repo(repo=Depends(RecipeRepository)):
    return repo

def get_post_repo(repo=Depends(PostRepository)):
    return repo

def get_users_repo(repo=Depends(AdminRepository)):
    return repo

def get_keywords(keywords=Depends(KeywordRepository)):
    return keywords

def get_bad_recipes(repo=Depends(BadRecipeRepository)):
    return repo
```

We can see from the preceding script that `RecipeRepository` is a dependable object of the factory methods, which are also injectable components but of the service layer. For instance, `get_recipe_repo()` will be wired to a service class to pursue the implementation of native services that require some transactions from `RecipeRepository`. In a way, we are indirectly wiring the repository class to the service layer.

The service layer

This layer has all the application's services with the domain logic, such as our `RecipeService`, which provides business processes and algorithms to `RecipeRepository`. The `get_recipe_repo()` factory is injected through its constructor to provide CRUD transactions from `RecipeRepository`. The injection strategy used here is the function of class dependency, which is depicted in the following code:

```python
from model.recipes import Recipe
from repository.factory import get_recipe_repo


class RecipeService:
    def __init__(self, repo=Depends(get_recipe_repo)):
        self.repo = repo

    def get_recipes(self):
        return self.repo.query_recipes()

    def add_recipe(self, recipe: Recipe):
        self.repo.insert_recipe(recipe)
```

The constructor of a typical Python class is always the appropriate place to inject components, which can either be a function or class dependable. With the preceding `RecipeService`, its `get_recipes()` and `add_recipe()` are realized because of the transactions derived from `get_recipe_repo()`.

The REST API and the service layer

The REST API methods can directly inject the service class or factory method if it needs to access the service layer. In our application, there is a factory method associated with each service class to apply the same strategy used in the `RecipeRepository` injection. That is why, in the following script, the `get_recipe_service()` method is wired to the REST API instead of `RecipeService`:

```python
class IngredientReq(BaseModel):
    id: UUID
```

```
        name:str
        qty: int
        measure: str

    class RecipeReq(BaseModel):
        id: UUID
        name: str
        ingredients: List[IngredientReq]
        cat: Category
        orig: Origin


router = APIRouter()

@router.post("/recipes/insert")
def insert_recipe(recipe: RecipeReq,
            handler=Depends(get_recipe_service)):
    json_dict = jsonable_encoder(recipe)
    rec = Recipe(**json_dict)
    handler.add_recipe(rec)
    return JSONResponse(content=json_dict, status_code=200)

@router.get("/recipes/list/all")
def get_all_recipes(handler=Depends(get_recipe_service)):
    return handler.get_recipes()
```

The insert_recipe() is a REST API that accepts a recipe and its ingredients from a client for persistency, while get_all_recipes() returns List[Recipe] as a response.

The actual project structure

With the power of DI, we have created an *online recipe system* with an organized set of *models*, *repository*, and *service* layers. The project structure shown in *Figure 3.3* is quite different from the previous prototypes because of the additional layers, but it still has main.py and all the packages and modules with their respective APIRouter.

Figure 3.3 – The Online Recipe System's project structure

At this point, DI has offered many advantages to FastAPI applications, from the engineering of object instantiation to breaking down monolithic components to set up loosely coupled structures. But there is only one slight problem: FastAPI's default container. The framework's container has no easy configuration to set all its managed objects to a *singleton* scope. Most applications prefer fetching singleton objects to avoid wasting memory in the **Python Virtual Machine** (**PVM**). Moreover, the built-in container is not open to a more detailed container configuration, such as having a *multiple container* setup. The next series of discussions will focus on the limitation of FastAPI's default container and solutions to overcome it.

Using third-party containers

DI has a lot to offer to improve our application, but it still depends on the framework we use to get the full potential of this design pattern. FastAPI's container is very acceptable to some when the concerns are simply on object management and project organization. However, when it comes to configuring the container to add more advanced features, it is not feasible for short-term projects, and it will be impossible for huge applications due to constraints. So, the practical way is to rely on *third-party modules* for the set of utilities needed to support all these advancements. So, let us explore these popular external modules that integrate seamlessly with FastAPI, the *Dependency Injector* and *Lagom*, which we can use to set up a complete and manageable container.

Using configurable containers – Dependency Injector

When it comes to configurable containers, the *Dependency Injector* has several module APIs that can be used to build variations of custom containers that can manage, assemble, and inject objects. But before we can use this module, we need to install it first using `pip`:

```
pip install dependency-injector
```

The containers and providers module

Among all the API types, *Dependency Injector* is popular with its *containers* and *providers*. One of its container types is `DeclarativeContainer`, which can be subclassed to contain all its providers. Its providers can be `Factory`, `Dict`, `List`, `Callable`, `Singleton`, or other *containers*. Both the `Dict` and `List` providers are easy to set up because they only need `list` and `dict` respectively to be instantiated. A `Factory` provider, conversely, instantiates any class, such as a repository, service, or a generic Python class, while `Singleton` only creates one instance per class, which is valid throughout the application's runtime. The `Callable` provider manages function dependencies, while `Container` instantiates other containers. Another container type is `DynamicContainer`, which is built from a configuration file, databases, or other resources.

The container types

Aside from these container APIs, the *Dependency Injector* allows us to customize a container based on the volume of the dependable objects, project structure, or other criteria from the project. The most common style or setup is the single declarative container that fits in small-, medium-, or large-scale applications. Our *online recipe system* prototype owns a single declarative container, which is implemented in the following script:

```python
from dependency_injector import containers, providers
from repository.users import login_details
from repository.login import LoginRepository
from repository.admin import AdminRepository
from repository.keywords import KeywordRepository
from service.recipe_utilities import get_recipe_names


class Container(containers.DeclarativeContainer):
    loginservice = providers.Factory(LoginRepository)
    adminservice = providers.Singleton(AdminRepository)
    keywordservice = providers.Factory(KeywordRepository)
    recipe_util = providers.Callable(get_recipe_names)
    login_repo = providers.Dict(login_details)
```

By simply subclassing `DeclarativeContainer`, we can easily create a single container, with its instances injected by the various providers previously mentioned. `LoginRepository` and `KeywordRepository` are both injected as new instances through the Factory provider. `AdminRepository` is an injected singleton object, `get_recipe_names()` is an injected function dependable, and `login_details` is an injected dictionary containing login credentials.

FastAPI and Dependency Injector integration

To wire the dependencies to a component through the Dependency Injector, the `@inject` decorator is applied. `@inject` is imported from the `dependency_injector.wiring` module and is decorated over the *dependent* component.

Afterward, the instance will be fetched from the container using the `Provide` wiring marker. Wiring markers search for the `Provider` object that references the injectable in the container, and if it exists, it will prepare for *auto-wiring*. Both `@inject` and `Provide` belong to the same API module:

```python
from repository.keywords import KeywordRepository
from containers.single_container import Container
from dependency_injector.wiring import inject, Provide
from uuid import UUID


router = APIRouter()


@router.post("/keyword/insert")
@inject
def insert_recipe_keywords(*keywords: str,
        keywordservice: KeywordRepository =
            Depends(Provide[Container.keywordservice])):
    if keywords != None:
        keywords_list = list(keywords)
        keywordservice.insert_keywords(keywords_list)
        return JSONResponse(content={"message":
           "inserted recipe keywords"}, status_code=201)
    else:
        return JSONResponse(content={"message":
           "invalid operation"}, status_code=403)
```

The integration happens when the `Depends()` function directive is invoked to register the wiring marker and the `Provider` instance to FastAPI. Aside from the acknowledgment, the registration adds *type hints* and *Pydantic validation rules* to the third-party `Provider` to appropriately wire the injectables into FastAPI. The preceding script imports `Container` from its module to wire `KeywordRepository` through `@inject`, the wire marker, and the `keywordservice` `Provider` of *Dependency Injector*.

Now, the last piece of the puzzle is to *assemble*, *create*, and *deploy* the single declarative container through the FastAPI platform. This last integration measure requires instantiating the *container* inside the module where the injections happened and then invoking its `wire()` method, which builds the assemblage. Since the preceding `insert_recipe_keywords()` is part of `/api/keywords.py`, we should add the following lines in the `keywords` module script, particularly at its end portion:

```
import sys

... ... ... ... ...

container = Container()
container.wire(modules=[sys.modules[__name__]])
```

The multiple container setup

For large applications, the number of repository transactions and services increases based on the functionality and special features of the application. If the single declarative type becomes unfeasible for a growing application, then we can always replace it with a *multiple-container* setup.

Dependency Injector allows us to create a separate container for each group of services. Our application has created a sample setup found in `/containers/multiple_containers.py`, just in case this prototype becomes full-blown. That sample of multiple declarative containers is shown as follows:

```
from dependency_injector import containers, providers

from repository.login import LoginRepository
from repository.admin import AdminRepository
from repository.keywords import KeywordRepository

class KeywordsContainer(containers.DeclarativeContainer):
    keywordservice = providers.Factory(KeywordRepository)

    ... ... ... ... ...


class AdminContainer(containers.DeclarativeContainer):
    adminservice = providers.Singleton(AdminRepository)

    ... ... ... ... ...


class LoginContainer(containers.DeclarativeContainer):
    loginservice = providers.Factory(LoginRepository)

    ... ... ... ... ...
```

```
class RecipeAppContainer(containers.DeclarativeContainer):
    keywordcontainer =
            providers.Container(KeywordsContainer)
    admincontainer = providers.Container(AdminContainer)
    logincontainer = providers.Container(LoginContainer)
    ... ... ... ... ...
```

Based on the preceding configuration, the three different instances of DeclarativeContainer created are KeywordsContainer, AdminContainer, and LoginContainer. The KeywordsContainer instance will assemble all dependencies related to keywords, AdminContainer will hold all instances related to administrative tasks, and LoginContainer for login- and user-related services. Then, there is RecipeAppContainer, which will consolidate all these containers through DI also.

The injection of the dependencies to the API is like the single declarative style, except that the container needs to be indicated in the wiring marker. The following is an admin-related API that shows how we wire dependencies to REST services:

```
from dependency_injector.wiring import inject, Provide

from repository.admin import AdminRepository
from containers.multiple_containers import
        RecipeAppContainer


router = APIRouter()


@router.get("/admin/logs/visitors/list")
@inject
def list_logs_visitors(adminservice: AdminRepository =
    Depends(
      Provide[
      RecipeAppContainer.admincontainer.adminservice])):
    logs_visitors_json = jsonable_encoder(
            adminservice.query_logs_visitor())
    return logs_visitors_json
```

The presence of admincontainer inside Provide checks first for the container of the same name before it fetches the adminservice provider that references the service dependable. The rest of the details are just the same with a single declarative, including the FastAPI integration and object assembly.

What is highlighted here about *Dependency Injector* is just basic configurations for simple applications. There are still other features and integrations that this module can provide to optimize our application using DI. Now, if we need thread-safe and non-blocking but with simple, streamlined, and straightforward APIs, setup and configuration, there is the *Lagom* module.

Using a simple configuration – Lagom

The third-party *Lagom* module is widely used because of its simplicity when it comes to wiring dependables. It is also ideal for building asynchronous microservice-driven applications because it is thread-safe at runtime. Moreover, it can easily integrate into many web frameworks, including FastAPI. To apply its APIs, we need to install it first using `pip`:

```
pip install lagom
```

The container

Containers in *Lagom* are created instantly using the `Container` class from its module. Unlike in *Dependency Injector*, Lagom's containers are created before the injection happens inside the module of the REST APIs:

```
from lagom import Container
from repository.complaints import BadRecipeRepository

container = Container()
container[BadRecipeRepository] = BadRecipeRepository()
router = APIRouter()
```

All dependables are injected into the container through typical instantiation. The container behaves like a `dict` when adding new dependables because it also uses a *key-value pair* as an entry. When we inject an object, the container needs its class name as its *key* and the instance as its *value*. Moreover, the DI framework also allows instantiation with arguments if the constructors require some parameter values.

The FastAPI and Lagom integration

Before the wiring happens, integration to the FastAPI platform must come first by instantiating a new API class called `FastApiIntegration`, which is found in the `lagom.integrations. fast_api` module. It takes `container` as a required parameter:

```
from lagom.integrations.fast_api import FastApiIntegration
deps = FastApiIntegration(container)
```

The dependables

The instance of FastAPIIntegration has a depends() method, which we will use to perform the injection. One of the best features of Lagom is its easy and seamless integration mechanism into any framework. Thus, wiring the dependencies will not need FastAPI's Depends() function anymore:

```
@router.post("/complaint/recipe")
def report_recipe(rid: UUID,
    complaintservice=deps.depends(BadRecipeRepository)):
        complaintservice.add_bad_recipe(rid)
        return JSONResponse(content={"message":
            "reported bad recipe"}, status_code=201)
```

The preceding report_recipe() utilizes BadRecipeRepository as an injectable service. Since it is part of the container, *Lagom*'s depends() function will search for the object in the container, and then it will be wired to the API service, if that exists, to save the complaints to the dict datastore.

So far, these two third-party modules are the most popular and elaborative when employing DI in our applications. These modules may change through future updates, but one thing is for sure: IoC and DI design patterns will always be the powerful solution in managing memory usage in an application. Let us now discusses issues surrounding memory space, container, and object assembly.

Scoping of dependables

In FastAPI, the scope of dependables can be either a new instance or a singleton. FastAPI's DI does not support the creation of singleton objects by default. In every execution of an API service with dependencies, FastAPI always fetches a new instance of each wired dependable, which can be proven by getting the *object ID* using id().

A singleton object is created only once by a container, no matter how many times the framework injects it. Its *object ID* remains the same the entire runtime of the application. Services and repository classes are preferred to be singleton to control the increase of memory utilization of the application. And since it is not easy to create a singleton with FastAPI, we can use either *Dependency Injector* or *Lagom*.

There is a `Singleton` provider in Dependency Injector that is responsible for the creation of singleton dependencies. This provider was mentioned already during the discussions on its `DeclarativeContainer` setup. With Lagom, there are two ways to create singleton injectables: (a) using its `Singleton` class, and (b) through the constructor of `FastAPIIntegration`.

The `Singleton` class wraps the instance of the dependency before injecting it into the container. The following sample snippet shows one example:

```
container = Container()
container[BadRecipeRepository] =
            Singleton(BadRecipeRepository())
```

The other way is to declare the dependency in the `request_singletons` parameter of the constructor of `FastAPIIntegration`. The following snippet shows how it is done:

```
container = Container()
container[BadRecipeRepository] = BadRecipeRepository()
deps = FastApiIntegration(container,
        request_singletons=[BadRecipeRepository])
```

By the way, the `request_singletons` parameter is a `List` type, so it will allow us to declare at least one dependable when we want to make singletons.

Summary

One aspect that makes a framework easy and practical to use is its support for the IoC principle. FastAPI has a built-in container that we can utilize to establish dependency among components. The use of a *DI* pattern to integrate all these components through wiring is a strong prerequisite in building microservice-driven applications. From simple injection using `Depends()`, we can extend DI to build pluggable components for database integration, authentication, security, and unit testing.

This chapter also introduced some third-party modules such as *Dependency Injector* and *Lagom* that can design and customize containers. Because of the limitations of FastAPI on DI, there are external libraries that can help extend its responsibility to assemble, control, and manage object creation in a container. These third-party APIs can also create singleton objects, which can help decrease the heap size in the PVM.

Aside from performance tuning and memory management, DI can contribute to the organization of a project, especially huge applications. The addition of model, repository, and service layers is a remarkable effect of creating dependencies. Injection opens the development to other design patterns, such as factory method, service, and data access object patterns. In the next chapter, we will start to build some microservice-related components based on the core design patterns of microservices.

4

Building the
Microservice Application

Previously, we spent a lot of time building API services for various applications using the core features of FastAPI. We also started applying important design patterns such as **Inversion of Control (IoC)** and **Dependency Injection (DI)**, which are essential for managing FastAPI container objects. External Python packages were installed and used to provide options on what containers to use in managing objects.

These design patterns can help not only with managed objects in container but also when building scalable, enterprise-grade, and unconventionally complex applications. Most of these design patterns help break down monolithic architecture into loosely coupled components that are known as *microservices*.

In this chapter, we will explore some architectural design patterns and principles that can provide strategies and ways to initiate the building of our microservices from a monolithic application. Our focus will be on breaking the huge application into business units, creating a sole gateway to bundle these business units, applying domain modeling to each of the microservices, and managing other concerns such as logging and application configuration.

Aside from expounding the benefits and disadvantages of each design pattern, another objective is to apply these architectural patterns to our software specimen to show its effectiveness and feasibility. And to support these goals, the following topics will be covered in this chapter:

- Applying the decomposition pattern
- Creating a common gateway
- Centralizing the logging mechanism
- Consuming the REST APIs
- Applying the domain modeling approach
- Managing a microservice's configuration details

Technical requirements

This chapter uses a *university ERP system* prototype that focuses on the students, faculty, and library submodules, but more on student-library and faculty-library operations (for example, book borrowing and issuing). Each submodule has its administration, management, and transaction services, and they are independent of each other even though they are part of an ERP specification. Currently, this sample prototype does not use any database management system, so all the data is temporarily stored in Python containers. The code is all uploaded at `https://github.com/PacktPublishing/` `Building-Python-Microservices-with-FastAPI` under the `ch04`, `ch04-student`, `ch04-faculty`, and `ch04-library` projects.

Applying the decomposition pattern

If we apply the monolithic strategy used in building the prototypes presented in the previous chapters, building this ERP will not be cost-effective in terms of resources and effort. There will be features that might become too dependent on other functions, which will put the teams of developers in a difficult situation whenever transaction problems occur due to these tightly coupled features. The best way to implement our University ERP prototype is to decompose the whole specification into smaller modules before the implementation starts.

There are two appropriate ways in which to decompose our application prototype, namely decomposition by business units and decomposition by subdomains:

- *Decomposition by business units* is used when the breakdown of the monolithic application is based on organizational structures, architectural components, and structural units. Usually, its resulting modules have fixed and structured processes and functionality that are seldom enhanced or upgraded.

- *Decomposition by subdomain* uses domain models and their corresponding business processes as the basis of the breakdown. Unlike the former, this decomposition strategy deals with modules that continuously evolve and change to capture the exact structure of the modules.

Of the two options, decomposition by business units is the more practical decomposition strategy to use for our monolithic University ERP prototype. Since the information and business flow used by universities has been part of its foundation for years, we need to organize and breakdown its voluminous and compounded operations by colleges or departments. *Figure 4.1* shows the derivation of these submodules:

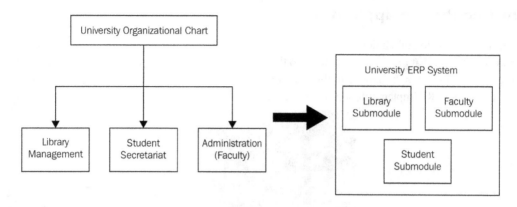

Figure 4.1 – Decomposition by business units

After determining the submodules, we can implement them as independent microservices using the FastAPI framework. We can call an implementation of a business unit or module a microservice if its services *can collectively stand as one component*. Also, it must be able to *collaborate with other microservices* through *interconnection* based on the URL address and port number. *Figure 4.2* shows the project directories of the faculty, library, and student management modules implemented as FastAPI microservice applications. *Chapter 1, Setting Up FastAPI for Starters,* to *Chapter 3, Investigating Dependency Injection,* gave us the foundation to build a FastAPI microservice:

Figure 4.2 – The faculty, library, and student microservice applications

Each of these microservices is independent of the others in terms of its server instance and management. Starting and shutting down one of them will not affect the other two, as each can have a different context root and port. Each application can have a separate logging mechanism, dependency environment, container, configuration file, and any other aspect of a microservice, which will be discussed in the subsequent chapters.

But FastAPI has another way of designing microservices using a *mount* sub-application.

Creating the sub-applications

FastAPI allows you to build independent sub-applications inside the *main application*. Here, `main.py` serves as a gateway that provides a pathname to these mounted applications. It also creates the mounts specifying the context path mapped to the FastAPI instance of each sub-application. *Figure 4.3* shows a new university ERP implementation that has been built using mounts:

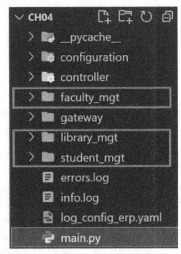

Figure 4.3 – The main project with the mounts

Here, `faculty_mgt`, `library_mgt`, and `student_mgt` are typical independent microservice applications mounted into the `main.py` component, the top-level application. Each sub-application has a `main.py` component, such as `library_mgt`, which has its FastAPI instance created in its `library_main.py` setup, as shown in the following code snippet:

```
from fastapi import FastAPI
library_app = FastAPI()
library_app.include_router(admin.router)
library_app.include_router(management.router)
```

The student sub-application has a `student_main.py` setup that creates its FastAPI instance, as shown in the following code:

```
from fastapi import FastAPI
student_app = FastAPI()
student_app.include_router(reservations.router)
student_app.include_router(admin.router)
```

```
student_app.include_router(assignments.router)
student_app.include_router(books.router)
```

Likewise, the faculty sub-application also has its `faculty_main.py` setup, as highlighted in the following code, for the same purpose, to build the microservice architecture:

```
from fastapi import FastAPI
faculty_app = FastAPI()
faculty_app.include_router(admin.router)
faculty_app.include_router(assignments.router)
faculty_app.include_router(books.router)
```

These sub-applications are typical FastAPI microservice applications containing all of the essential components such as routers, middleware exception handlers, and all the necessary packages to build REST API services. The only difference from the usual applications is that their context paths or URLs are defined and decided by the top-level application that handles them.

> **Important note**
>
> Optionally, we can run the `library_mgt` sub-application independently from `main.py` through the `uvicorn main:library_app --port 8001` command, `faculty_mgt` through `uvicorn main:faculty_app --port 8082`, and `student_mgt` through `uvicorn main:student_app --port 8003`. The option of running them independently despite the mount explains why these mounted sub-applications are all microservices.

Mounting the submodules

All the FastAPI decorators of each sub-application must be mounted in the `main.py` component of the top-level application for them to be accessed at runtime. The `mount()` function is invoked by the FastAPI decorator object of the top-level application, which adds all FastAPI instances of the sub-applications into the gateway application (`main.py`) and maps each with its corresponding URL context. The following script shows how the mounting of the *library*, *student*, and *faculty* subsystems is implemented in the `main.py` component of the University ERP top-level system:

```
from fastapi import FastAPI
from student_mgt import student_main
from faculty_mgt import faculty_main
from library_mgt import library_main


app = FastAPI()
```

```
app.mount("/ch04/student", student_main.student_app)
app.mount("/ch04/faculty", faculty_main.faculty_app)
app.mount("/ch04/library", library_main.library_app)
```

With this setup, the mounted /ch04/student URL will be used to access all the API services of the *student module* app, /ch04/faculty will be used for all the services of the *faculty module*, and /ch04/library will be used for the *library*-related REST services. These mounted paths become valid once they are declared in mount() because FastAPI automatically handles all of these paths through the root_path specification.

Since all three sub-applications of our *university ERP system* are independent microservices, now let us apply another design strategy that can help manage the requests to these applications just by using the main URL of the ERP system. Let us utilize the *main application* as a gateway to our sub-applications.

Creating a common gateway

It will be easier if we use the URL of the main application to manage the requests and redirect users to any of the three sub-applications. The *main application* can stand as a pseudo-reverse proxy or an entry point for user requests, which will always redirect user requests to any of the desired sub-applications. This kind of approach is based on a design pattern called *API Gateway*. Now, let us explore how we can apply this design to manage independent microservices mounted onto the main application using a workaround.

Implementing the main endpoint

There are so many solutions when it comes to implementing this gateway endpoint, and among them is having a simple REST API service in the top-level application with an integer path parameter that will identify the ID parameter of the microservice. If the ID parameter is invalid, the endpoint will only return the {'message': 'University ERP Systems'} JSON string instead of an error. The following script is a straightforward implementation of this endpoint:

```
from fastapi import APIRouter


router = APIRouter()


@router.get("/university/{portal_id}")
def access_portal(portal_id:int):
    return {'message': 'University ERP Systems'}
```

The `access_portal` API endpoint is created as a GET path operation with `portal_id` as its path parameter. The `portal_id` parameter is essential to this process because it will determine which among the *Student*, *Faculty*, and *Library* microservices the user wants to access. Therefore, accessing the `/ch04/university/1` URL should lead the user to the student application, `/ch04/university/2` to the faculty microservice, and `/ch04/university/3` to the library application.

Evaluating the microservice ID

The `portal_id` parameter will automatically be fetched and evaluated using a dependable function that is injected into the `APIRouter` instance where the API endpoint is implemented. As discussed in *Chapter 3, Investigating Dependency Injection*, a *dependable function* or *object* can serve as a filter or validator of all incoming requests of any services once injected into an `APIRouter` or `FastAPI` instance. The dependable function used in this ERP prototype, as shown in the following script, evaluates whether the `portal_id` parameter is 1, 2, or 3 only:

```
def call_api_gateway(request: Request):
    portal_id = request.path_params['portal_id']
    print(request.path_params)
    if portal_id == str(1):
        raise RedirectStudentPortalException()
    elif portal_id == str(2):
        raise RedirectFacultyPortalException()
    elif portal_id == str(3):
        raise RedirectLibraryPortalException()
class RedirectStudentPortalException(Exception):
    pass


class RedirectFacultyPortalException(Exception):
    pass


class RedirectLibraryPortalException(Exception):
    pass
```

The given solution is a feasible workaround to trigger a custom event since FastAPI has no built-in event handling except for the startup and shutdown event handlers, which are topics in *Chapter 8, Creating Coroutines, Events, and Message-Driven Transactions*. So, once `call_api_gateway()` finds `portal_id` to be a valid microservice ID, it will raise some custom exceptions. It will throw `RedirectStudentPortalException` if the user wants to access the *Student* microservice. On the other hand, the `RedirectFacultyPortalException` error will be raised if the user desires the *Faculty* microservice. Otherwise, the `RedirectLibraryPortalException` error will be triggered when the *Library* microservice is the one wanted by the user. But first, we need to inject `call_api_gateway()` into the `APIRouter` instance handling the gateway endpoint through the `main.py` component of the top-level ERP application. The following script shows you how it is injected into `university.router` using the concepts discussed earlier:

```
from fastapi import FastAPI, Depends, Request, Response
from gateway.api_router import call_api_gateway
from controller import university


app = FastAPI()
app.include_router (university.router,
          dependencies=[Depends(call_api_gateway)],
          prefix='/ch04')
```

All of these raised exceptions require an exception handler that will listen to the throws and execute some of the tasks required to pursue the microservices.

Applying the exception handlers

The exception handler does a redirection to the appropriate microservice. As you learned in *Chapter 2, Exploring the Core Features*, each thrown exception must have its corresponding exception handler to pursue the required response after the exception handling. Here are the exception handlers that will handle the custom exception thrown by `call_api_gateway()`:

```
from fastapi.responses import RedirectResponse
from gateway.api_router import call_api_gateway,
        RedirectStudentPortalException,
        RedirectFacultyPortalException,
        RedirectLibraryPortalException
```

```python
@app.exception_handler(RedirectStudentPortalException)
def exception_handler_student(request: Request,
    exc: RedirectStudentPortalException) -> Response:
    return RedirectResponse(
        url='http://localhost:8000/ch04/student/index')


@app.exception_handler(RedirectFacultyPortalException)
def exception_handler_faculty(request: Request,
    exc: RedirectFacultyPortalException) -> Response:
    return RedirectResponse(
        url='http://localhost:8000/ch04/faculty/index')


@app.exception_handler(RedirectLibraryPortalException)
def exception_handler_library(request: Request,
    exc: RedirectLibraryPortalException) -> Response:
    return RedirectResponse(
        url='http://localhost:8000/ch04/library/index')
```

Here, exception_handler_student() will redirect the user to the mount path of the *Student* microservice, while exception_handler_faculty() will redirect the user to the *Faculty* sub-application. Additionally, exception_handler_library() will let the user access the *Library* microservice. Exception handlers are the last component needed to complete the API Gateway architecture. The exceptions trigger the redirection to the independent microservices mounted on the FastAPI framework.

Although there are other, better solutions to achieve the gateway architecture, our approach is still procedural and pragmatic without having to resort to external modules and tools, just the core components of FastAPI. *Chapter 11, Adding Other Microservices Features*, will discuss establishing an effective API Gateway architecture using Docker and NGINX.

Now, let us explore how to set up a centralized logging mechanism for this kind of microservices setup.

Centralizing the logging mechanism

We have created an audit trail mechanism with middleware and Python file transactions in *Chapter 2, Exploring the Core Features*. We have found out that middleware, which can only be set up through the FastAPI decorator of the top-level application, can manage incoming Request and outgoing Response of any API services. This time, we will be using custom middleware to set up a centralized logging feature that will log all service transactions of the top-level application alongside its independent mounted microservices. Of the many approaches for integrating these logging concerns into the application without changing the API services, we will concentrate on the following pragmatic custom approach with the custom middleware and *Loguru* module.

Utilizing the Loguru module

An *application log* is essential to any enterprise-grade application. For monolithic applications deployed in a single server, logging means letting service transactions write their log messages to a single file. On the other hand, logging can be too complex and complicated to implement in an independent microservices setup, especially when these services are for deployment to different servers or Docker containers. Its logging mechanism could even cause runtime problems if the module used is not adaptable to asynchronous services.

For FastAPI instances that support both asynchronous and synchronous API services that run on an ASGI server, using Python's logging module always generates the following error log:

```
2021-11-08 01:17:22,336 - uvicorn.error - ERROR - Exception in
ASGI application
Traceback (most recent call last):
  File "c:\alibata\development\language\python\
  python38\lib\site-packages\uvicorn\protocols\http\
  httptools_impl.py", line 371, in run_asgi
    result = await app(self.scope, self.receive, self.send)
  File "c:\alibata\development\language\python\
  python38\lib\site-packages\uvicorn\middleware\
  proxy_headers.py", line 59, in __call__
    return await self.app(scope, receive, send)
```

Opting for another logging extension is the only solution to avoid the error generated by the `logging` module. The best option is one that can fully support the FastAPI framework, which is the `loguru` extension. But first, we need to install it using the `pip` command:

```
pip install loguru
```

Loguru is a straightforward and easy-to-use logging extension. We can immediately log using its default handler, the `sys.stderr` handler, even without adding much configurations. Since our application needs to place all messages in a log file, we need to add the following lines to the `main.py` component of the top-level application right after the instantiation of `FastAPI`:

```
from loguru import logger
from uuid import uuid4


app = FastAPI()
app.include_router (university.router,
        dependencies=[Depends(call_api_gateway)],
        prefix='/ch04')
logger.add("info.log",format="Log: [{extra[log_id]}:
    {time} - {level} - {message} ", level="INFO",
    enqueue = True)
```

Note that its `logger` instance has an `add()` method where we can register *sinks*. The first part of the *sinks* is the *handler* that decides whether to emit the logs in `sys.stdout` or the file. In our university ERP prototype, we need to have a global `info.log` file that contains all the log messages of the sub-applications.

A crucial part of the log sink is the `level` type, which indicates the granularity of log messages that need to be managed and logged. If we set the `level` parameter of `add()` to `INFO`, it tells the logger to consider only those messages under the `INFO`, `SUCCESS`, `WARNING`, `ERROR`, and `CRITICAL` weights. The logger will bypass log messages outside these levels.

Another part of the *sinks* is the `format` log, where we can create a custom log message layout to replace its default format. This format is just like a Python interpolated string without the *"f"* that contains placeholders such as `{time}`, `{level}`, `{message}`, and any custom placeholders that need to be replaced by `logger` at runtime.

In `log.file`, we want our logs to start with the `Log` keyword followed immediately by the custom-generated `log_id` parameter and then the time the logging happened, the level, and the message.

And to add support for asynchronous logging, the `add()` function has an `enqueue` parameter that we can enable anytime. In our case, this parameter is default to `True` just to prepare for any `async/await` execution.

There is a lot to explore with Loguru's features and functionality. For instance, we can create additional handlers for loggers to emit where each of these handlers has different retention, rotation, and rendition types. Additionally, Loguru can allow us to add colors to our logs through some color markups such as <red>, <blue>, or <cyan>. It also has an @catch() decorator that can be applied to manage exceptions at runtime. All the logging features we need to set up our unified application log are in Loguru. Now that we have configured our Loguru in the top-level application, we need to let its logging mechanism work across the three sub-applications or microservices without modifying their code.

Building the logging middleware

The core component of this centralized application log is the custom middleware that we must implement in the main.py component where we set up Loguru. FastAPI's *mount* allows us to centralize some cross-cutting concerns such as logging without adding anything to the sub-applications. One middleware implementation in the main.py component of the top-level application is good enough to pursue logging across the independent microservices. The following is the middleware implementation for our specimen application:

```
@app.middleware("http")
async def log_middleware(request:Request, call_next):
    log_id = str(uuid4())
    with logger.contextualize(log_id=log_id):
        logger.info('Request to access ' +
            request.url.path)
        try:
            response = await call_next(request)
        except Exception as ex:
            logger.error(f"Request to " +
                request.url.path + " failed: {ex}")
            response = JSONResponse(content=
                {"success": False}, status_code=500)
        finally:
            logger.info('Successfully accessed ' +
                request.url.path)
            return response
```

First, `log_middleware()` will generate a `log_id` parameter every time it intercepts any API services from the main app or the sub-applications. Then, the `log_id` parameter is injected into the `dict` of context information through Loguru's `contextualize()` method since `log_id` is part of the log information, as indicated in our log format setup. Afterward, logging starts before the API service is executed and after its successful execution. When exceptions are encountered during the process, the logger will still generate a log message with the `Exception` message. So, whenever we access any API services anywhere from the ERP prototype, the following log messages will be written in `info.log`:

```
Log: [1e320914-d166-4f5e-a39b-09723e04400d: 2021-11-
28T12:02:25.582056+0800 - INFO - Request to access /ch04/
university/1
Log: [1e320914-d166-4f5e-a39b-09723e04400d: 2021-11-
28T12:02:25.597036+0800 - INFO - Successfully accessed /ch04/
university/1
Log: [fd3badeb-8d38-4aec-b2cb-017da853e3db: 2021-11-
28T12:02:25.609162+0800 - INFO - Request to access /ch04/
student/index
Log: [fd3badeb-8d38-4aec-b2cb-017da853e3db: 2021-11-
28T12:02:25.617177+0800 - INFO - Successfully accessed /ch04/
student/index
Log: [4cdb1a46-59c8-4762-8b4b-291041a95788: 2021-11-
28T12:03:25.187495+0800 - INFO - Request to access /ch04/
student/profile/add
Log: [4cdb1a46-59c8-4762-8b4b-291041a95788: 2021-11-
28T12:03:25.203421+0800 -
INFO - Request to access /ch04/faculty/index
Log: [5cde7503-cb5e-4bda-aebe-4103b2894ffe: 2021-11-
28T12:03:33.432919+0800 - INFO - Successfully accessed /ch04/
faculty/index
Log: [7d237742-fdac-4f4f-9604-ce49d3c4c3a7: 2021-11-
28T12:04:46.126516+0800 - INFO - Request to access /ch04/
faculty/books/request/list
Log: [3a496d87-566c-477b-898c-8191ed6adc05: 2021-11-
28T12:04:48.212197+0800 - INFO - Request to access /ch04/
library/book/request/list
Log: [3a496d87-566c-477b-898c-8191ed6adc05: 2021-11-
28T12:04:48.221832+0800 - INFO - Successfully accessed /ch04/
library/book/request/list
Log: [7d237742-fdac-4f4f-9604-ce49d3c4c3a7: 2021-11-
28T12:04:48.239817+0800 -
```

```
Log: [c72f4287-f269-4b21-a96e-f8891e0a4a51: 2021-11-
28T12:05:28.987578+0800 - INFO - Request to access /ch04/
library/book/add
Log: [c72f4287-f269-4b21-a96e-f8891e0a4a51: 2021-11-
28T12:05:28.996538+0800 - INFO - Successfully accessed /ch04/
library/book/add
```

The given snapshot of log messages proves that we have a centralized setup because the middleware filters all API service execution and performs the logging transaction. It shows that the logging started from accessing the gateway down to executing the API services from the *faculty*, *student*, and *library* sub-applications. Centralizing and managing cross-cutting concerns is one advantage that can be provided by using FastAPI's *mounting* when building independent microservices.

But when it comes to the interactions among these independent sub-applications, can mounting also be an advantage? Now, let us explore how independent microservices in our architecture can communicate by utilizing each other's API resources.

Consuming the REST API services

Just like in an unmounted microservices setup, mounted ones can also communicate by accessing each other's API services. For instance, if a faculty member or student wants to borrow a book from the library, how can that setup be implemented seamlessly?

In *Figure 4.4*, we can see that interactions can be possible by establishing a client-server communication wherein one API service can serve as a resource provider, and the others are the clients:

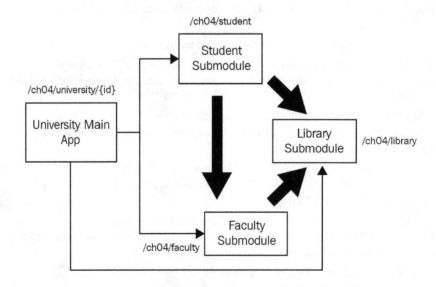

Figure 4.4 – Interaction with the faculty, student, and library microservices

Consuming API resources in FastAPI can be straightforward using the `httpx` and `requests` external modules. The following discussions will focus on how these two modules can help our mounted services interact with each other.

Using the httpx module

The `httpx` external module is a Python extension that can consume both asynchronous and synchronous REST APIs and has *HTTP/1.1* and *HTTP/2* support. It is a fast and multi-purpose toolkit that is used to access API services running on WSGI-based platforms, as well as, on ASGI, like the FastAPI services. But first, we need to install it using `pip`:

```
pip install httpx
```

Then, we can use it directly without further configuration to make two microservices interact, for instance, our *student* module submitting assignments to the *faculty* module:

```python
import httpx


@router.get('/assignments/list')
async def list_assignments():
    async with httpx.AsyncClient() as client:
     response = await client.get(
      "http://localhost:8000/ch04/faculty/assignments/list")
     return response.json()


@router.post('/assignment/submit')
def submit_assignment(assignment:AssignmentRequest ):
    with httpx.Client() as client:
        response = client.post("http://localhost:8000/
            ch04/faculty/assignments/student/submit",
             data=json.dumps(jsonable_encoder(assignment)))
        return response.content
```

The `httpx` module can process the GET, POST, PATCH, PUT, and DELETE path operations. It can allow the passing of different request parameters to the requested API without so much complexity. The `post()` client operation, for instance, can accept headers, cookies, params, json, files, and model data as parameter values. We use the `with` context manager to directly manage the streams created by its `Client()` or `AsyncClient()` instances, which are closeable components.

The preceding `list_assignments` service is a client that uses the `AsyncClient()` instance to pursue its GET request from an asynchronous `/ch04/faculty/assignments/list` API endpoint from the *faculty* module. `AsyncClient` accesses the WSGI-based platform to execute any asynchronous services, not the synchronous ones, or else it will throw *Status Code 500*. It might require additional configuration details in its constructor for some complex cases, where it needs to further manage resource access through ASGI.

On the other hand, the `submit_assignment` service is a synchronous client that accesses another synchronous endpoint, `ch04/faculty/assignments/student/submit`, which is a POST HTTP operation. In this case, the `Client()` instance is used to access the resource to submit an assignment to the *Faculty* module through a POST request. `AssignmentRequest` is a `BaseModel` object that needs to be filled up by the client for submission to the request endpoint. Unlike `params` and `json`, which are passed straightforwardly as `dict`, `data` is a model object that must be first converted into `dict` by `jsonable_encoder()` and `json.dumps()` to make the transport feasible across the HTTP. The new converted model becomes the argument value of the `data` parameter of the POST client operation.

When it comes to the response of the client services, we can allow the response to be treated as a piece of text using the module's `content` or as a JSON result using `json()`. It now depends on the requirement of the client service as to what response type to use for the application.

Using the requests module

Another option to establish client-server communication among microservices is the `requests` module. Although `httpx` and `requests` are almost compatible, the latter offers other features such as auto-redirection and explicit session handling. The only problem with `requests` is its non-direct support to asynchronous APIs and its slow performance when accessing resources. Despite its drawbacks, the `requests` module is still the standard way of consuming REST APIs in Python microservice development. First, we need to install it before we can use it:

```
pip install requests
```

In our ERP prototype, the `requests` extension was used by the *faculty* microservice to borrow books from the *library* module. Let's look at the *Faculty* client services that show us how the `requests` module is used to access the synchronous API of *library*:

```
@router.get('/books/request/list')
def list_all_request():
    with requests.Session() as sess:
        response = sess.get('http://localhost:8000/
            ch04/library/book/request/list')
        return response.json()
```

```python
@router.post('/books/request/borrow')
def request_borrow_book(request:BookRequestReq):
    with requests.Session() as sess:
        response = sess.post('http://localhost:8000/
            ch04/library/book/request',
                data=dumps(jsonable_encoder(request)))
        return response.content


@router.get('/books/issuance/list')
def list_all_issuance():
    with requests.Session() as sess:
        response = sess.get('http://localhost:8000/
            ch04/library/book/issuance/list')
        return response.json()


@router.post('/books/returning')
def return_book(returning: BookReturnReq):
    with requests.Session() as sess:
        response = sess.post('http://localhost:8000/
            ch04/library/book/issuance/return',
                data=dumps(jsonable_encoder(returning)))
        return response.json()
```

The requests module has a Session() instance, which is equivalent to Client() in the httpx module. It provides all the necessary client operations that will consume the API endpoints from the FastAPI platform. Since Session is a closeable object, the context manager is, again, used here to handle the streams that will be utilized during the access of the resources and transport of some parameter values. Like in httpx, parameter details such as params, json, header, cookies, files, and data are also part of the requests module and are ready for transport through the client operation if needed by the API endpoints.

From the preceding code, we can see that sessions are created to implement the list_all_request and list_all_issuance GET client services. Here, request_borrow_book is a POST client service that requests a book in the form of BookRequestReq from the /ch04/library/book/request API endpoint. Similar to httpx, jsonable_encoder() and json.dumps() must be used to convert the BaseModel object into dict in order to be transported as a data parameter value. The same approach is also applied to the return_book POST client service, which returns the book borrowed by the faculty. The responses of these client services can also be content or json() just like what we have in the httpx extension.

Using the `requests` and `httpx` modules allows these mounted microservices to interact with each other based on some specification. Consuming exposed endpoints from other microservices minimizes tight coupling and strengthens the importance of the decomposition design pattern of building independent microservices.

The next technique gives you the option of managing components within a microservice using *domain modeling*.

Applying the domain modeling approach

Applications that are database-focused or built from core functionalities without collaborating with models are either not easy to manage when they scale up or not friendly when enhanced or bug-fixed. The reason behind this is the absence of the structure and flow of business logic to follow, study, and analyze. Understanding the behavior of an application and deriving the domain models with the business logic behind them encompasses the best approach when it comes to establishing and organizing a structure in an application. This principle is called the *domain modeling approach*, which we will now apply to our ERP specimen.

Creating the layers

Layering is one implementation that is inevitable when applying domain-driven development. There is a dependency between *layers* that, sometimes, might pose problems when fixing bugs during development. But what are important in layered architectures are the concepts, structures, categories, functionalities, and roles that *layering* can create, which helps in understanding the specification of the application. *Figure 4.5* shows the *models*, *repositories*, *services*, and *controllers* of the sub-applications:

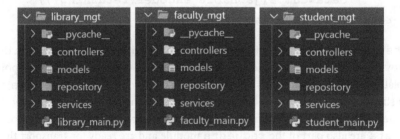

Figure 4.5 – Layered architecture

The most crucial layer is the `models` layer, which consists of the domain model classes that describe the domain and business processes involved in the application.

Identifying the domain models

The *domain model layer* is the initial artifact of the application because it provides the contextual framework of the application. Business processes and transactions can be easily classified and managed if domains are first determined during the initial phase of the development. The code organization created by domain layering can provide code traceability, which can ease source code updates and debugging.

In our ERP specimen, these models are categorized into two: the *data* and *request* models. The data models are those used to capture and store data in its temporary data stores, while the request models are the BaseModel objects used in the API services.

For instance, the *faculty* module has the following data models:

```python
class Assignment:
    def __init__(self, assgn_id:int, title:str,
        date_due:datetime, course:str):
        self.assgn_id:int = assgn_id
        self.title:str = title
        self.date_completed:datetime = None
        self.date_due:datetime = date_due
        self.rating:float = 0.0
        self.course:str = course

    def __repr__(self):
      return ' '.join([str(self.assgn_id), self.title,
        self.date_completed.strftime("%m/%d/%Y, %H:%M:%S"),
        self.date_due.strftime("%m/%d/%Y, %H:%M:%S"),
        str(self.rating) ])

    def __expr__(self):
      return ' '.join([str(self.assgn_id), self.title,
       self.date_completed.strftime("%m/%d/%Y, %H:%M:%S"),
       self.date_due.strftime("%m/%d/%Y, %H:%M:%S"),
       str(self.rating) ])

class StudentBin:
    def __init__(self, bin_id:int, stud_id:int,
      faculty_id:int):
```

```
        self.bin_id:int = bin_id
        self.stud_id:int = stud_id
        self.faculty_id:int = faculty_id
        self.assignment:List[Assignment] = list()


    def __repr__(self):
        return ' '.join([str(self.bin_id),
          str(self.stud_id), str(self.faculty_id)])


    def __expr__(self):
        return ' '.join([str(self.bin_id),
          str(self.stud_id), str(self.faculty_id)])
```

These data model classes always have their constructors implemented if constructor injection is needed during instantiation. Moreover, the __repr__() and __str__() dunder methods are optionally there to provide efficiency for developers when accessing, reading, and logging these objects.

On the other hand, the request models are familiar because they were already discussed in the previous chapter. Additionally, the *faculty* module has the following request models:

```
class SignupReq(BaseModel):
    faculty_id:int
    username:str
    password:str


class FacultyReq(BaseModel):
    faculty_id:int
    fname:str
    lname:str
    mname:str
    age:int
    major:Major
    department:str


class FacultyDetails(BaseModel):
    fname:Optional[str] = None
    lname:Optional[str] = None
```

```
mname:Optional[str] = None
age:Optional[int] = None
major:Optional[Major] = None
department:Optional[str] = None
```

The request models listed in the preceding snippet are just simple `BaseModel` types. For further details on how to create `BaseModel` classes, *Chapter 1*, *Setting Up FastAPI for Starters*, provides guidelines for creating different kinds of `BaseModel` classes to capture different requests from clients.

Building the repository and service layers

The two most popular domain modeling patterns that are crucial in building the layers of this approach are the *repository* and *service layer patterns*. The repository aims to create strategies for managing data access. Some repository layers only provide data connectivity to the data store like in our specimen here, but oftentimes, repository's goal is to interact with the **Object Relational Model (ORM)** framework to optimize and manage data transactions. But aside from the access, this layer provides a high-level abstraction for the application so that the specific database technology or *dialect* used will not matter to the applications. It serves as an adapter to any database platform to pursue data transactions for the application, nothing else. The following is a repository class of the *faculty* module, which manages the domain for creating assignments for their students:

```python
from fastapi.encoders import jsonable_encoder
from typing import List, Dict, Any
from faculty_mgt.models.data.facultydb import
    faculty_assignments_tbl
from faculty_mgt.models.data.faculty import Assignment
from collections import namedtuple


class AssignmentRepository:

    def insert_assignment(self,
            assignment:Assignment) -> bool:
        try:
            faculty_assignments_tbl[assignment.assgn_id] =
                assignment
        except:
            return False
        return True
```

```python
def update_assignment(self, assgn_id:int,
        details:Dict[str, Any]) -> bool:
    try:
        assignment = faculty_assignments_tbl[assgn_id]
        assignment_enc = jsonable_encoder(assignment)
        assignment_dict = dict(assignment_enc)
        assignment_dict.update(details)
        faculty_assignments_tbl[assgn_id] =
        Assignment(**assignment_dict)
    except:
        return False
    return True


def delete_assignment(self, assgn_id:int) -> bool:
    try:
        del faculty_assignments_tbl[assgn_id]
    except:
        return False
    return True


def get_all_assignment(self):
    return faculty_assignments_tbl
```

Here, `AssignmentRepository` manages the `Assignment` domain object using its four repository transactions. Additionally, `insert_assignment()` creates a new `Assignment` entry in the `faculty_assignment_tbl` dictionary, and `update_assignment()` accepts new details or the corrected information of an existing assignment and updates it. On the other hand, `delete_assignment()` deletes an existing `Assignment` entry from the data store using its `assign_id` parameter. To retrieve all the created assignments, the repository class has `get_all_assignment()`, which returns all the entries of `faculty_assignments_tbl`.

The service layer pattern defines the algorithms, operations, and process flows of the applications. Oftentimes, it interacts with the repository to build the necessary business logic, management, and controls for the other components of the application, such as the API services or controllers. Usually, one service caters to one repository class or more depending on the specification of the project. The following code snippet is a service that interfaces a repository to provide additional tasks such as UUID generation for a student workbin:

```python
from typing import List, Dict , Any
from faculty_mgt.repository.assignments import
            AssignmentSubmissionRepository
from faculty_mgt.models.data.faculty import Assignment
from uuid import uuid4


class AssignmentSubmissionService:

    def __init__(self):
        self.repo:AssignmentSubmissionRepository =
            AssignmentSubmissionRepository()

    def create_workbin(self, stud_id:int, faculty_id:int):
        bin_id = uuid4().int
        result = self.repo.create_bin(stud_id, bin_id,
                    faculty_id )
        return (result, bin_id)

    def add_assigment(self, bin_id:int,
                    assignment: Assignment):
        result = self.repo.insert_submission(bin_id,
                    assignment )
        return result

    def remove_assignment(self, bin_id:int,
                    assignment: Assignment):
        result = self.repo.insert_submission(bin_id,
                    assignment )
        return result
```

```
    def list_assignments(self, bin_id:int):
        return self.repo.get_submissions(bin_id)
```

The `AssignmentSubmissionService` cited in the preceding code has methods that utilize the `AssignmentSubmissionRepository` transactions. It provides them with parameters and returns the `bool` results for evaluation by other components. Other services might look more complicated than this sample because algorithms and tasks are usually added to pursue the requirements of the layers.

The successful wiring of a repository class to the service happens in the latter's constructor. Usually, the repository class is instantiated just like in the preceding sample. Another fantastic option is to use DI, as discussed in *Chapter 3*.

Using the factory method pattern

The *factory method design pattern* is always a good approach for managing injectable classes and functions using the `Depends()` component. *Chapter 3* showcased factory methods as mediums to inject the repository components into the service instead of instantiating them directly within the service. The design pattern provides loose coupling between components or layers. This approach is highly applicable to large applications wherein some modules and sub-components are reused and inherited.

Now, let us look at how the top-level application can manage the different configuration details of these mounted and independent microservice applications.

Managing a microservice's configuration details

So far, this chapter has provided us with some popular design patterns and strategies that can give us a kickstart on how to provide our FastAPI microservices with the best structures and architecture. This time, let us explore how the FastAPI framework supports storing, assigning, and reading configuration details to mounted microservice applications such as database credentials, networking configuration data, application server information, and deployment details. First, we need to install `python-dotenv` using `pip`:

```
pip install python-dotenv
```

All of these settings are values that are external to the implementation of the microservice applications. Instead of hardcoding them into the code as variable data, usually, we store them in the *env*, *property*, or *INI* files. However, challenges arise when assigning these settings to different microservices.

Frameworks that support the *externalized configuration design pattern* have an internal processing feature that fetches environment variables or settings without requiring additional parsing or decoding techniques. For instance, the FastAPI framework has built-in support for externalized settings through pydantic's `BaseSettings` class.

Storing settings as class attributes

In our architecture setup, it should be the top-level application that will manage the externalized values. One way is to store them in a `BaseSettings` class as attributes. The following are classes of the `BaseSettings` type with their respective application details:

```
from pydantic import BaseSettings
from datetime import date


class FacultySettings(BaseSettings):
    application:str = 'Faculty Management System'
    webmaster:str = 'sjctrags@university.com'
    created:date = '2021-11-10'


class LibrarySettings(BaseSettings):
    application:str = 'Library Management System'
    webmaster:str = 'sjctrags@university.com'
    created:date = '2021-11-10'


class StudentSettings(BaseSettings):
    application:str = 'Student Management System'
    webmaster:str = 'sjctrags@university.com'
    created:date = '2021-11-10'
```

Here, `FacultySettings` will be assigned to the *faculty* module since it carries some information regarding the module. `LibrarySettings` is for the *library* module to utilize, while `StudentSettings` is for the *student* module.

To fetch the values, first, a component in a module must import the `BaseSettings` class from the main project's `/configuration/config.py` module. Then, it needs an injectable function to instantiate it before injecting it into a component that needs to utilize the values. The following script is part of `/student_mgt/student_main.py`, where the settings need to be retrieved:

```
from configuration.config import StudentSettings
```

```
student_app = FastAPI()
student_app.include_router(reservations.router)
student_app.include_router(admin.router)
student_app.include_router(assignments.router)
student_app.include_router(books.router)


def build_config():
    return StudentSettings()


@student_app.get('/index')
def index_student(
    config:StudentSettings = Depends(build_config)):
    return {
        'project_name': config.application,
        'webmaster': config.webmaster,
        'created': config.created
    }
```

Here, build_config() is an injectable function that injects the StudentSettings instance into the /index endpoint of the *student* microservice. After the DI, the application, webmaster, and created values will become accessible from the config wired object. These settings will appear on the browser right after calling the /ch04/university/1 gateway URL.

Storing settings in the properties file

Another option is to store all these settings inside a physical file with an extension of .env, .properties, or .ini. For instance, this project has the erp_settings.properties file found in the /configuration folder, and it contains the following application server details in *key-value* pair format:

```
production_server = prodserver100
prod_port = 9000
development_server = devserver200
dev_port = 10000
```

To fetch these details, the application needs another `BaseSettings` class implementation that declares the *key* of the *key-value* pair as attributes. The following class shows how `production_server`, `prod_port`, `development_server`, and `dev_port` are declared without any assigned values:

```
import os

class ServerSettings(BaseSettings):
    production_server:str
    prod_port:int
    development_server:str
    dev_port:int

    class Config:
        env_file = os.getcwd() +
            '/configuration/erp_settings.properties'
```

Aside from the class variable declaration, `BaseSetting` requires an implementation of an *inner class*, called `Config`, with a predefined `env_file` assigned to the current location of the properties file.

The same processes are involved when it comes to accessing the property details from the file. After importing `ServerSettings`, it needs an injectable function to inject its instance to the components that need the details. The following script is an updated version of `/student_mgt/student_main.py`, which includes access to the `development_server` and `development_port` settings:

```
from fastapi import FastAPI, Depends
from configuration.config import StudentSettings,
      ServerSettings

student_app = FastAPI()
student_app.include_router(reservations.router)
student_app.include_router(admin.router)
student_app.include_router(assignments.router)
student_app.include_router(books.router)

def build_config():
    return StudentSettings()

def fetch_config():
    return ServerSettings()
```

```
@student_app.get('/index')
def index_student(
    config:StudentSettings = Depends(build_config),
    fconfig:ServerSettings = Depends(fetch_config)):
    return {
        'project_name': config.application,
        'webmaster': config.webmaster,
        'created': config.created,
        'development_server' : fconfig.development_server,
        'dev_port': fconfig.dev_port
    }
```

Based on this enhanced script, running the /ch04/university/1 URL will redirect the browser to a screen showing additional server details from the properties file. Managing configuration details in FastAPI is easy, as we either save them inside a class or inside a file. No external module is necessary, and no special coding effort is required to fetch all these settings, just the creation of the BaseSettings classes. This easy setup contributes to building flexible and adaptable microservice applications that can run on varying configuration details.

Summary

The chapter started with the decomposition pattern, which is useful for breaking down a monolithic application into granularized, independent, and scalable modules. The FastAPI application that implemented these modules exhibited some principles included in the *12-Factor Application principles* of a microservice, such as having independence, configuration files, logging systems, code bases, port binding, concurrency, and easy deployment.

Alongside decomposition, this chapter also showcased the *mounting of different independent sub-applications* onto the FastAPI platform. Only FastAPI can group independent microservices using mounts and bind them into one port with their corresponding context roots. From this feature, we created a pseudo-API Gateway pattern that serves as a façade to the independent sub-applications.

Despite the possible drawbacks, the chapter also highlighted domain modeling as an option for organizing components in a FastAPI microservice. The *domain*, *repository*, and *service* layers help manage the information flow and task distribution based on the project specification. Tracing, testing, and debugging are easy when domain layers are in place.

In the next chapter, we will focus on integrating our microservice applications with a relational database platform. The focus is to establish database connectivity and utilize our data models to implement CRUD transactions within the repository layer.

Part 2: Data-Centric and Communication-Focused Microservices Concerns and Issues

In this part of the book, we will be exploring other FastAPI components and features to solve other design patterns that the API framework can build, looking at data, communication, messaging, reliability, and security. External modules will also be highlighted in order to pursue other behavior and frameworks, such as ORM and reactive programming.

This part comprises the following chapters:

- *Chapter 5, Connecting to a Relational Database*
- *Chapter 6, Using a Non-Relational Database*
- *Chapter 7, Securing the REST APIs*
- *Chapter 8, Creating Coroutines, Events, and Message-Driven Transactions*

5

Connecting to a Relational Database

Our previous applications have only used Python collections to hold data records instead of persistent data stores. This setup causes data wiping whenever the **Uvicorn** server restarts because these collections only store the data in *volatile memory*, such as **RAM**. From this chapter onward, we will be applying data persistency to avoid data loss and provide a platform to manage our records, even when the server is in shutdown mode.

This chapter will focus on different **Object Relational Mappers (ORMs)** that can efficiently manage clients' data using objects and a relational database. Object-relational mapping is a technique where SQL statements for **Creating, Reading, Updating** *and* **Deleting (CRUD)** are implemented and executed in an object-oriented programming approach. ORM requires all relationships or tables to be mapped to their corresponding entity or model classes to avoid tightly coupled connections to the database platform. And these model classes are the ones that are used to connect to the database.

Aside from introducing ORM, this chapter will also discuss a design pattern called **Command and Query Responsibility Segregation (CQRS)**, which can help resolve conflicts between read and write ORM transactions at the domain level. CQRS can help minimize the running time spent by read and write SQL transactions to improve the overall performance of the application over time compared to the data modeling approach.

Overall, the main objective of this chapter is to prove that the FastAPI framework supports all popular ORMs to provide applications with backend database access, which it does by using popular relational database management systems, and apply optimization to CRUD transactions using the CQRS design pattern.

In this chapter, we will cover the following topics:

- Preparing for database connectivity
- Creating synchronous CRUD transactions using *SQLAlchemy*
- Implementing asynchronous CRUD transactions using *SQLAlchemy*
- Using *GINO* for asynchronous CRUD transactions
- Using Pony ORM for the repository layer
- Building the repository using Peewee
- Applying the CQRS design pattern

Technical requirements

The application prototype that's been created for this chapter is called *fitness club management system*; it caters to membership and gym fitness operations. This prototype has administration, membership, class management, and attendance modules that utilize a **PostgreSQL** database as their data storage. Moreover, this uncommon application has four pieces of database connectivity that have been configured using different ORM variations to provide you with options for your applications. Also, the prototype is just a simple FastAPI application that's been created to help you focus on the *data modeling features*, *data persistency*, and *query building* required for the discussions in this chapter. This code for this chapter can be found at `https://github.com/PacktPublishing/Building-Python-Microservices-with-FastAPI` in the `ch05a` and `ch05b` projects.

Preparing for database connectivity

Let us consider some application-related concerns before we start discussing database connectivity in FastAPI:

- First, all the application prototypes from this chapter onward will be using PostgreSQL as the sole relational DBMS. We can download its installer from `https://www.enterprisedb.com/downloads/postgres-postgresql-downloads`.
- Second, the *fitness club management system* prototype has an existing database called `fcms` with six tables, namely `signup`, `login`, `profile_members`, `profile_trainers`, `attendance_member`, and `gym_class`. All these tables, along with their metadata and relationships, can be seen in the following diagram:

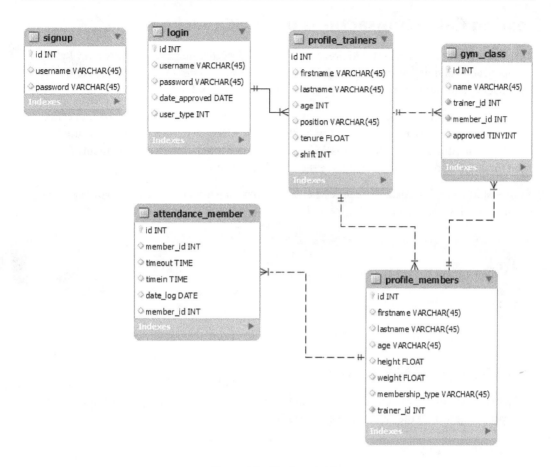

Figure 5.1 – The fcms tables

The project folder contains a script called `fcms_postgres.sql` that installs all these schemas.

Now that we've installed the latest version of PostgreSQL and run the fcms script file, let us learn about SQLAlchemy, the most widely used ORM library in the Python arena.

> **Important note**
>
> This chapter will compare and contrast the features of different Python ORMs. With this experimental setup, each project will have a multitude of database connectivity, which is against the convention of having a single piece of database connectivity per project.

Creating CRUD transactions using SQLAlchemy

SQLAlchemy is the most popular ORM library and can establish communication between any Python-based application and database platform. It is reliable because it is continuously updated and tested to be efficient, high-performing, and accurate with its SQL reads and writes.

This ORM is a boilerplated interface that aims to create a database-agnostic data layer that can connect to any database engine. But compared to other ORMs, SQLAlchemy is DBA-friendly because it can generate optimized native SQL statements. When it comes to formulating its queries, it only requires Python functions and expressions to pursue the CRUD operations.

Before we start using SQLAlchemy, check whether you have the module installed in your system by using the following command:

```
pip list
```

If SQLAlchemy is not in the list, install it using the `pip` command:

```
pip install SQLAlchemy
```

Currently, the version being used while developing the *fitness club management system* app is *1.4*.

Installing the database driver

SQLAlchemy will not work without the required database driver. It is mandatory to install the `psycopg2` dialect since the database of choice is PostgreSQL:

```
pip install psycopg2
```

Psycopg2 is a *DB API 2.0-*compliant PostgreSQL driver that does connection pooling and can work with multi-threaded FastAPI applications. This wrapper or dialect is also essential in building synchronous CRUD transactions for our application. Once it's been installed, we can start looking at SQLAlchemy's database configuration details. All the code related to SQLAlchemy can be found in the `ch05a` project.

Setting up the database connection

To connect to any database, SQLAlchemy requires an engine that manages the connection pooling and the installed dialect. The `create_engine()` function from the `sqlalchemy` module is the source of the engine object. But to successfully derive it, `create_engine()` requires a database URL string to be configured. This URL string contains the *database name, the database API driver*, the *account credentials*, the *IP address* of the database server, and its *port*. The following script shows how to create the engine that will be used in our *fitness club management system* prototype:

```
from sqlalchemy import create_engine
```

```
DB_URL =
    "postgresql://postgres:admin2255@localhost:5433/fcms"

engine = create_engine(DB_URL)
```

engine is a global object and must be created only once in the entire application. Its first database connection happens right after the first SQL transaction of the application because it follows the *lazy initialization* design pattern.

Moreover, engine in the previous script is essential for creating the ORM session that will be used by SQLAlchemy to execute CRUD transactions.

Initializing the session factory

All CRUD transactions in SQLAlchemy are driven by *sessions*. Each session manages a group of database "writes" and "reads," and it checks whether to execute them or not. For instance, it maintains a group of inserted, updated, and deleted objects, checks whether the changes are valid, and then coordinates with the SQLAlchemy core to pursue the changes to the database if all transactions have been validated. It follows the behavior of the *Unit of Work* design pattern. SQLAlchemy relies on sessions for data consistency and integrity.

But before we create a session, the data layer needs a session factory that is bound to the derived engine. The ORM has a sessionmaker() directive from the sqlalchemy.orm module, which requires the engine object. The following script shows how to invoke sessionmaker():

```
from sqlalchemy.orm import sessionmaker

engine = create_engine(DB_URL)
SessionFactory = sessionmaker(autocommit=False,
                    autoflush=False, bind=engine)
```

Apart from engine binding, we also need to set the session's autocommit property to False to impose commit() and rollback() transactions. The application should be the one to flush all changes to the database, so we need to set its autoflush feature to False as well.

Applications can create more than one session through the SessionFactory() call, but having one session per APIRouter is recommended.

Defining the Base class

Next, we need to set up the `Base` class, which is crucial in mapping model classes to database tables. Although SQLAlchemy can create tables at runtime, we opted to utilize an existing schema definition for our prototype. Now, this `Base` class must be subclassed by the model classes so that the mapping to the tables will happen once the server starts. The following script shows how straightforward it is to set up this component:

```
from sqlalchemy.ext.declarative import declarative_base


Base = declarative_base()
```

Invoking the `declarative_base()` function is the easiest way of creating the Base class rather than creating `registry()` to call `generate_base()`, which can also provide us with the Base class.

Note that all these configurations are part of the `/db_config/sqlalchemy_connect.py` module of the prototype. They are bundled into one module since they are crucial in building the SQLAlchemy repository. But before we implement the CRUD transactions, we need to create the model layer using the `Base` class.

Building the model layer

The model classes of SQLAlchemy have all been placed in the `/models/data/sqlalchemy_models.py` file of the fitness club project folder. If `BaseModel` is important to API request models, the Base class is essential in building the data layer. It is imported from the configuration file to define SQLAlchemy entities or models. The following code is from the module script, which shows how we can create model class definitions in SQLAlchemy ORM:

```python
from sqlalchemy import Time, Boolean, Column, Integer,
    String, Float, Date, ForeignKey
from sqlalchemy.orm import relationship
from db_config.sqlalchemy_connect import Base


class Signup(Base):
    __tablename__ = "signup"


    id = Column(Integer, primary_key=True, index=True)
    username = Column('username', String, unique=False,
                        index=False)
    password = Column('password' ,String, unique=False,
                        index=False)
```

The Signup class is a sample of a SQLAlchemy model because it inherits the Base class's properties. It is a mapped class because all its attributes are reflections of the column metadata of its physical table schema counterpart. The model has a primary_key property set to True because SQLAlchemy recommends each table schema have at least one primary key. The rest of the Column objects are mapped to column metadata that's non-primary but can be *unique* or *indexed*. Each model class inherits the __tablename__ property, which sets the name of the mapped table.

Most importantly, we need to ensure that the data type of the class attribute matches the column type of its column counterpart in the table schema. The column attribute must have the same name as the column counterpart. Otherwise, we need to specify the actual column name in the first argument of the Column class, as shown in the username and password columns of Signup. But most of the time, we must always make sure they are the same to avoid confusion.

Mapping table relationships

SQLAlchemy strongly supports different types of parent-child or associative table relationships. Model classes involved in the relationship require the relationship() directive from the sqlalchemy. orm module to be utilized to establish one-to-many or one-to-one relationships among model classes. This directive creates a reference from the parent to the child class using some foreign key indicated in the table schema definition.

A child model class uses the ForeignKey construct in its foreign key column object to link the model class to its parent's reference key column object. This directive indicates that the values in this column should be within the values stored in the parent table's reference column. The ForeignKey directive applies to both the primary and non-primary Column objects. The following model class defines a sample column relationship in our database schema:

```python
class Login(Base):
    __tablename__ = "login"

    id = Column(Integer, primary_key=True, index=True)
    username = Column(String, unique=False, index=False)
    password = Column(String, unique=False, index=False)
    date_approved = Column(Date, unique=False, index=False)
    user_type = Column(Integer, unique=False, index=False)

    trainers = relationship('Profile_Trainers',
        back_populates="login", uselist=False)
    members = relationship('Profile_Members',
        back_populates="login", uselist=False)
```

This `Login` model is linked to two children, `Profile_Trainers` and `Profile_Members`, based on its configuration. Both child models have the `ForeignKey` directive in their `id` column objects, as shown in the following model definitions:

```
class Profile_Trainers(Base):
    __tablename__ = "profile_trainers"
    id = Column(Integer, ForeignKey('login.id'),
        primary_key=True, index=True, )
    firstname = Column(String, unique=False, index=False)
    ... ... ... ... ...
    ... ... ... ... ...
    login = relationship('Login',
        back_populates="trainers")
    gclass = relationship('Gym_Class',
        back_populates="trainers")

class Profile_Members(Base):
    __tablename__ = "profile_members"
    id = Column(Integer, ForeignKey('login.id'),
        primary_key=True, index=True)
    firstname = Column(String, unique=False, index=False)
    lastname = Column(String, unique=False, index=False)
    age = Column(Integer, unique=False, index=False)
    ... ... ... ... ... ...
    ... ... ... ... ... ...
    trainer_id = Column(Integer,
        ForeignKey('profile_trainers.id'), unique=False,
        index=False)

    login = relationship('Login', back_populates="members")
    attendance = relationship('Attendance_Member',
        back_populates="members")
    gclass = relationship('Gym_Class',
        back_populates="members")
```

The `relationship()` directive is the sole directive for creating table relationships. We need to specify some of its parameters, such as the *name of the child model class* and the *backreference specification*. The `back_populates` parameter refers to the complementary attribute names of the related model classes. This indicates the rows that need to be fetched using some relationship loading technique during join query transactions. The `backref` parameter can also be used instead of `back_populates`.

On the other hand, `relationship()` can return either a `List` or scalar object, depending on the relationship type. If it is a *one-to-one type*, the parent class should set the `useList` parameter to `False` to indicate that it will return a scalar value. Otherwise, it will select a list of records from the child table. The previous `Login` class definition shows that `Profile_Trainers` and `Profile_Members` hold a one-to-one relationship with `Login` because `Login` sets its `uselist` to `False`. On the other hand, the model relationship between `Profile_Members` and `Attendance_Member` is a *one-to-many* type because `uselist` is set to `True` by default, as shown by the following definitions:

```
class Attendance_Member(Base):
    __tablename__ = "attendance_member"
    id = Column(Integer, primary_key=True, index=True)
    member_id = Column(Integer,
        ForeignKey('profile_members.id'), unique=False,
        index=False)
    timeout = Column(Time, unique=False, index=False)
    timein = Column(Time, unique=False, index=False)
    date_log = Column(Date, unique=False, index=False)


    members = relationship('Profile_Members',
            back_populates="attendance")
```

While setting the model relationships, we must also consider the *relationship loading type* that these related model classes will be using during the join query transactions. We specify this detail in the `lazy` parameter of `relationship()`, which is assigned to `select` by default. This is because SQLAlchemy uses a lazy loading technique by default in retrieving join queries. However, you can modify it to use `joined` (`lazy="joined"`), subquery (`lazy="subquery"`), select in (`lazy="selectin"`), raise (`lazy="raise"`), or no (`lazy="no"`) loading. Among the options, the `joined` approach is better for *INNER JOIN* transactions.

Implementing the repository layer

In the SQLAlchemy ORM, creating the repository layer requires the *model classes* and a `Session` object. The `Session` object, derived from the `SessionFactory()` directive, establishes all the communication to the database and manages all the model objects before the `commit()` or `rollback()` transaction. When it comes to the queries, the `Session` entity stores the result set of records in a data structure called an *identity map*, which maintains the unique identity of each data record using the primary keys.

All repository transactions are *stateless*, which means the session is automatically closed after loading the model objects for insert, update, and delete transactions when the database issues a `commit()` or `rollback()` operation. We import the `Session` class from the `sqlalchemy.orm` module.

Building the CRUD transactions

Now, we can start building the repository layer of the fitness club application since we have already satisfied the requirements to build the CRUD transactions. The following `SignupRepository` class is the blueprint that will show us how to *insert*, *update*, *delete*, and *retrieve* record(s) to/from the `signup` table:

```python
from typing import Dict, List, Any
from sqlalchemy.orm import Session
from models.data.sqlalchemy_models import Signup
from sqlalchemy import desc


class SignupRepository:

    def __init__(self, sess:Session):
        self.sess:Session = sess

    def insert_signup(self, signup: Signup) -> bool:
        try:
            self.sess.add(signup)
            self.sess.commit()
        except:
            return False
        return True
```

So far, `insert_signup()` is the most accurate way of persisting records to the `signup` table using SQLAlchemy. `Session` has an `add()` method, which we can invoke to add all record objects to the table, and a `commit()` transaction to finally flush all the new records into the database. The `flush()` method of `Session` is sometimes used instead of `commit()` to pursue the insertion and close `Session`, but most developers often use the latter. Note that the `signup` table contains all the gym members and trainers who want to gain access to the system. Now, the next script implements update record transaction:

```python
def update_signup(self, id:int,
        details:Dict[str, Any]) -> bool:
    try:
        self.sess.query(Signup).
            filter(Signup.id == id).update(details)
        self.sess.commit()
    except:
        return False
    return True
```

The `update_signup()` provides a short, straightforward, and robust solution to updating a record in SQLAlchemy. Another possible solution is to query the record through `self.sess.query(Signup).filter(Signup.id == id).first()`, replace the attribute values of the retrieved object with the new values from the `details` dictionary, and then invoke `commit()`. This way is acceptable, but it takes three steps rather than calling the `update()` method after `filter()`, which only takes one. Next script is an implementation of a delete record transaction:

```python
def delete_signup(self, id:int) -> bool:
    try:
        signup = self.sess.query(Signup).
                filter(Signup.id == id).delete()
        self.sess.commit()
    except:
        return False
    return True
```

On the other hand, `delete_signup()` follows the strategy of `update_signup()`, which uses `filter()` first before `delete()` is called. Another way of implementing this is to retrieve the object using `sess.query()` again and pass the retrieved object as an argument to the `Session` object's `delete(obj)`, which is a different function. Always remember to invoke `commit()` to flush the changes. Now, the following script shows how to implement the query transactions:

```
def get_all_signup(self):
    return self.sess.query(Signup).all()

def get_all_signup_where(self, username:str):
    return self.sess.
        query(Signup.username, Signup.password).
        filter(Signup.username == username).all()

def get_all_signup_sorted_desc(self):
    return self.sess.
        query(Signup.username, Signup.password).
        order_by(desc(Signup.username)).all()

def get_signup(self, id:int):
    return self.sess.query(Signup).
        filter(Signup.id == id).one_or_none()
```

Moreover, `SignupRepository` also highlights multiple and single records being retrieved in many forms. The `Session` object has a `query()` method, which requires *model class(es)* or *model column names* as argument(s). The function argument performs the record retrieval with column projection. For instance, the given `get_all_signup()` selects all signup records with all the columns projected in the result. If we want to include only `username` and `password`, we can write our query as `sess.query(Signup.username, Signup.password)`, just like in the given `get_all_signup_where()`. This `query()` method also shows how to manage constraints using the `filter()` method with the appropriate conditional expressions. Filtering always comes after column projection.

On the other hand, the `Session` object has an `order_by()` method that takes column names as parameters. It is performed last in the series of query transactions, before the result is extracted. The given sample, `get_all_signup_sorted_desc()`, sorts all `Signup` objects in descending order by `username`.

The last portion of the `query()` builder returns the result of the transactions, whether it is a list of records or a single record. The `all()` function ends the query statement that returns multiple records, while `first()`, `scalar()`, `one()`, or `one_or_none()` can be applied if the result is a single row. In `get_signup()`, `one_or_none()` is utilized to raise an exception when no record is returned. For SQLAlchemy's query transactions, all these functions can close the `Session` object. The repository classes for SQLAlchemy are in the ch05a folder's `/repository/sqlalchemy/signup.py` module script file.

Creating the JOIN queries

For all the ORMs supported by FastAPI, only SQLAlchemy implements join queries pragmatically and functionally, just like how we implemented the previous CRUD transactions. We used almost all of the methods we need to create joins previously except for `join()`.

Let us look at `LoginMemberRepository`, which shows how we can create a join query statement in SQLAlchemy with model classes in *one-to-one relationships*:

```
class LoginMemberRepository():
    def __init__(self, sess:Session):
        self.sess:Session = sess


    def join_login_members(self):
        return self.sess.
            query(Login, Profile_Members).
                filter(Login.id == Profile_Members.id).all()
```

`join_login_members()` shows the conventional way of creating *JOIN* queries. This solution requires passing the parent and child classes as query parameters and overriding the ON condition through the `filter()` method. The parent model class must come first in the column projection before the child class in the `query()` builder to extract the preferred result.

Another way is to use the `select_from()` function instead of `query()` to distinguish the parent class from the child. This approach is more appropriate for a *one-to-one* relationship.

On the other hand, `MemberAttendanceRepository` showcases the *one-to-many* relationship between the `Profile_Members` and `Attendance_Member` model classes:

```
class MemberAttendanceRepository():
    def __init__(self, sess:Session):
        self.sess:Session = sess


    def join_member_attendance(self):
```

```
        return self.sess.
            query(Profile_Members, Attendance_Member).
            join(Attendance_Member).all()

    def outer_join_member(self):
        return self.sess.
            query(Profile_Members, Attendance_Member).
            outerjoin(Attendance_Member).all()
```

join_member_attendance() shows the use of the join() method in building the *INNER JOIN* queries between Profile_Members and Attendance_Member. filter() is not needed anymore to build the ON condition because join() automatically detects and recognizes the relationship() parameters and the ForeignKey constructs defined at the beginning. But if there are other additional constraints, filter() can always be invoked, but only after the join() method.

The outer_join_member() repository method implements an *OUTER JOIN* query from the one-to-many relationship. The outerjoin() method will extract all Profile_Members records mapped to their corresponding Attendance_Member or return null if there are none.

Running the transactions

Now, let us apply these repository transactions to the administration-related API services of our application. Instead of using collections to store all the records, we will be utilizing the ORM's transactions to manage the data using PostgreSQL. First, we need to import the essential components required by the repository, such as SessionFactory, the repository class, and the Signup model class. APIs such as Session and other typing APIs can only be part of the implementation for type hints.

The following script shows a portion of the administrator's API services highlighting the insertion and retrieval services for new access registration:

```
from fastapi import APIRouter, Depends
from fastapi.responses import JSONResponse

from sqlalchemy.orm import Session
from db_config.sqlalchemy_connect import SessionFactory
from repository.sqlalchemy.signup import SignupRepository,
    LoginMemberRepository, MemberAttendanceRepository
from typing import List

router = APIRouter()
```

```
def sess_db():
    db = SessionFactory()
    try:
        yield db
    finally:
        db.close()
```

First, we need to create the Session instance through SessionFactory(), which we derived from sessionmaker(), since the repository layer is dependent on the session. In our application, a sess_db() custom generator is used to open and destroy the Session instance. It is injected into the API service methods to tell the Session instance to proceed with instantiating SignupRepository:

```
@router.post("/signup/add")
def add_signup(req: SignupReq,
        sess:Session = Depends(sess_db)):
    repo:SignupRepository = SignupRepository(sess)
    signup = Signup(password= req.password,
                username=req.username,id=req.id)
    result = repo.insert_signup(signup)
    if result == True:
        return signup
    else:
        return JSONResponse(content={'message':'create
                signup problem encountered'},
            status_code=500)
```

Once instantiated, the repository can provide record insertion through insert_signup(), which inserts the Signup record. Another of its methods is get_all_signup(), which retrieves all login accounts for approval:

```
@router.get("/signup/list", response_model=List[SignupReq])
def list_signup(sess:Session = Depends(sess_db)):
    repo:SignupRepository = SignupRepository(sess)
    result = repo.get_all_signup()
    return result


@router.get("/signup/list/{id}", response_model=SignupReq)
def get_signup(id:int, sess:Session = Depends(create_db)):
```

```
repo:SignupRepository = SignupRepository(sess)
result = repo.get_signup(id)
return result
```

Both the `get_signup()` and `list_signup()` services have a `request_model` of the `SignupReq` type, which determines the expected output of the APIs. But as you may have noticed, `get_signup()` returns the `Signup` object, while `list_signup()` returns a list of `Signup` records. How is that possible? If `request_model` is used to capture the query result of the SQLAlchemy query transactions, the `BaseModel` class or request model must include a nested `Config` class with its `orm_mode` set to `True`. This built-in configuration enables type mapping and validation of `BaseModel` for the SQLAlchemy model types used by the repository, before all the record objects are filtered and stored in the request models. More information about the `response_model` parameter can be found in *Chapter 1, Setting Up FastAPI for Starters*.

`SignupReq`, which is used by the query services of our application, is defined as follows:

```python
from pydantic import BaseModel


class SignupReq(BaseModel):
    id : int
    username: str
    password: str


    class Config:
        orm_mode = True
```

The script shows how `orm_mode` is enabled using the equals sign (=) rather than the typical colon symbol (:), which means `orm_mode` is a configuration detail and not part of the class attribute.

Overall, using SQLAlchemy for the repository layer is systematic and procedural. It is easy to map and synchronize the model classes with the schema definitions. Establishing relationships through the model classes is handy and predictable. Although there are lots of APIs and directives involved, it is still the most widely supported library for domain modeling and repository construction. Its documentation (`https://docs.sqlalchemy.org/en/14/`) is complete and informative enough to guide developers regarding the different API classes and methods.

Another feature of SQLAlchemy that's loved by many is its capability to generate table schemas at the application level.

Creating tables

Usually, SQLAlchemy works with the table schemas that have already been generated by the database administrator. In this project, the ORM setup started with designing the domain model classes before mapping them to the actual tables. But SQLAlchemy can auto-create table schemas at runtime for the FastAPI platform, which may be helpful during the testing or prototyping stage of the project.

The `sqlalchemy` module has a `Table()` directive that can create a table object with the essential column metadata using the `Column()` method, which we used in the mapping. The following is a sample script that shows how the ORM creates the signup table at the application level:

```
from sqlalchemy import Table, Column, Integer, String,
                MetaData
from db_config.sqlalchemy_connect import engine
meta = MetaData()

signup = Table(
    'signup', meta,
    Column('id', Integer, primary_key = True,
            nullable=False),
    Column('username', String, unique = False,
            nullable = False),
    Column('password', String, unique = False,
            nullable = False),
)
meta.create_all(bind=engine)
```

Part of the schema definition is `MetaData()`, a registry that contains the necessary methods for generating the tables. When all the schema definitions are signed off, the `create_all()` method of the `MetaData()` instance is executed with the engine to create the tables. This process may sound straightforward, but we seldom pursue this DDL feature of SQLAlchemy in projects at the production stage.

Now, let us explore how SQLAlchemy can be used to create asynchronous CRUD transactions for asynchronous API services.

Implementing async CRUD transactions using SQLAlchemy

From version 1.4, SQLAlchemy supports **asynchronous I/O (AsyncIO)** features, which enables support for asynchronous connections, sessions, transactions, and database drivers. Most of the procedures for creating the repository are the same as those for the synchronous setup. The only difference is the non-direct access that the CRUD commands have with the asynchronous Session object. Our ch05b project showcases the asynchronous side of SQLAlchemy.

Installing the asyncio-compliant database drivers

Before we begin setting up the database configuration, we need to install the following asyncio-compliant drivers: aiopg and asyncpg. First, we need to install aiopg, a library that will assist with any asynchronous access to PostgreSQL:

```
pip install aiopg
```

Next, we must install asyncpg, which helps build PostgreSQL asynchronous transactions through Python's AsyncIO framework:

```
pip install asyncpg
```

This driver is a *non-database API-compliant* driver because it runs on top of the AsyncIO environment instead of the database API specification for synchronous database transactions.

Setting up the database's connection

After installing the necessary drivers, we can derive the database engine through the application's create_async_engine() method, which creates an asynchronous version of SQLAlchemy's Engine, known as AsyncEngine. This method has parameters to set such as future, which can enable a variety of asynchronous features during CRUD transactions when set to True. Also, it has an echo parameter that can provide us with the generated SQL queries in the server log at runtime. But the most essential is the database URL, which now reflects the asynchronous database access through calling the asyncpg protocol. The following is the complete script for the asynchronous connection to the PostgreSQL database:

```
from sqlalchemy.ext.asyncio import create_async_engine

DB_URL =
    "postgresql+asyncpg://postgres:admin2255@
        localhost:5433/fcms"
```

```
engine = create_async_engine(DB_URL, future=True,
                echo=True)
```

The additional `"+asyncpg"` detail in DB_URL indicates that psycopg2 will no longer be the core database driver for PostgreSQL; instead, asyncpg will be used. This detail enables AsyncEngine to utilize asyncpg to establish the connection to the database. Omitting this detail will instruct the engine to recognize the psycopg2 database API driver, which will cause problems during the CRUD transactions.

Creating the session factory

Like in the synchronous version, the sessionmaker() directive is utilized to create the session factory with some new parameters set to enable AsyncSession. First, its expire_on_commit parameter is set to False to make that model instances and its attribute values accessible for the duration of the transaction, even after calling commit(). Unlike in the synchronous environment, all entity classes and their column objects are still accessible by other processes, even after transaction commit. Then, its class_ parameter bears the class name AsyncSession, the entity that will take control of the CRUD transactions. Of course, sessionmaker() still needs the engine for AsyncConnection and its underlying asynchronous context managers.

The following script shows how the session factory is derived using the sessionmaker() directive:

```
from sqlalchemy.ext.asyncio import AsyncSession
from sqlalchemy.orm import sessionmaker

engine = create_async_engine(DB_URL, future=True,
                echo=True)
AsynSessionFactory = sessionmaker(engine,
        expire_on_commit=False, class_=AsyncSession)
```

The full configuration for the asynchronous SQLAlchemy database connection can be found in the /db_config/sqlalchemy_async_connect.py module script file. Let us now create the model layer.

Creating the Base class and the model layer

Creating the Base class using declarative_base() and creating the model classes using Base is the same as what we did in the synchronous version. No additional parameters are needed to build the data layer for the asynchronous repository transactions.

Building the repository layer

Implementing asynchronous CRUD transactions is entirely different from implementing synchronous ones. The ORM supports the use of the `execute()` method of the `AsyncConnection` API to run some of the built-in ORM core methods, namely `update()`, `delete()`, and `insert()`. When it comes to query transactions, the new `select()` directive from the `sqlalchemy.future` module is used instead of the core `select()` method. And since `execute()` is an `async` method, this requires that all repository transactions are `async` too to apply the *Async/Await* design pattern. The following `AttendanceRepository` uses the asynchronous type of SQLAlchemy:

```python
from typing import List, Dict, Any
from sqlalchemy import update, delete, insert
from sqlalchemy.future import select
from sqlalchemy.orm import Session
from models.data.sqlalchemy_async_models import
            Attendance_Member


class AttendanceRepository:

    def __init__(self, sess:Session):
        self.sess:Session = sess


    async def insert_attendance(self, attendance:
            Attendance_Member) -> bool:
        try:
            sql = insert(Attendance_Member).
                values(id=attendance.id,
                    member_id=attendance.member_id,
                    timein=attendance.timein,
                    timeout=attendance.timeout,
                    date_log=attendance.date_log)
            sql.execution_options(
                    synchronize_session="fetch")
            await self.sess.execute(sql)
        except:
            return False
        return True
```

The given *asynchronous* `insert_attendance()` method in the preceding script shows the use of the `insert()` directive in creating an attendance log for a gym member. First, we need to pass the model class name to `insert()` to let the session know what table to access for the transaction. Afterward, it emits the `values()` method to project all the column values for insertion. Lastly, we need to call the `execute()` method to run the final `insert()` statement and automatically commit the changes since we didn't turn off the `autocommit` parameter of `sessionmaker()` during the configuration. Do not forget to invoke `await` before running the asynchronous method because everything runs on top of the AsyncIO platform this time.

Also, you have the option to add some additional execution details before running `execute()`. One of these options is `synchronize_session`, which tells the session to always synchronize the model attribute values and the updated values from the database using the `fetch` method.

Almost the same procedure is applied to the `update_attendance()` and `delete_attendance()` methods. We can run them through `execute()` and nothing else:

```python
async def update_attendance(self, id:int,
        details:Dict[str, Any]) -> bool:
    try:
        sql = update(Attendance_Member).where(
            Attendance_Member.id == id).values(**details)
        sql.execution_options(
            synchronize_session="fetch")
        await self.sess.execute(sql)

    except:
        return False
    return True


async def delete_attendance(self, id:int) -> bool:
    try:
        sql = delete(Attendance_Member).where(
            Attendance_Member.id == id)
        sql.execution_options(
            synchronize_session="fetch")
        await self.sess.execute(sql)
    except:
        return False
    return True
```

When it comes to queries, the repository class contains get_all_attendance(), which retrieves all the attendance records, and get_attendance(), which retrieves the attendance log of a particular member through its id. Constructing the select() method is a straightforward and pragmatic task since it is similar to writing a native SELECT statement in SQL development. First, the method needs to know what columns to project, and then it caters to some constraints if there are any. Then, it needs the execute() method to run the query asynchronously and extract the Query object. The resulting Query object has a scalars() method, which we can call to retrieve the list of records. Do not forget to close the session by calling the all() method.

check_attendance(), on the other hand, uses the scalar() method of the Query object to retrieve one record: a specific attendance. Aside from record retrieval, scalar() also closes the session:

```python
async def get_all_attendance(self):
    q = await self.sess.execute(
            select(Attendance_Member))
    return q.scalars().all()


async def get_attendance(self, id:int):
    q = await self.sess.execute(
        select(Attendance_Member).
            where(Attendance_Member.member_id == id))
    return q.scalars().all()


async def check_attendance(self, id:int):
    q = await self.sess.execute(
        select(Attendance_Member).
            where(Attendance_Member.id == id))
    return q.scalar()
```

The repository classes for the asynchronous SQLAlchemy can be found in the /repository/sqlalchemy/attendance.py module script file. Now, let us apply these asynchronous transactions to pursue some attendance monitoring services for our fitness gym application.

> **Important note**
> The ** operator in update_attendance() is a Python operator overload that converts a dictionary into kwargs. Thus, the result of **details is a kwargs argument for the values() method of the select() directive.

Running the CRUD transactions

There two big differences between AsyncIO-driven SQLAlchemy and the database API-compliant option when creating the `Session` instance:

- First, `AsyncSession`, which was created by the `AsyncSessionFactory()` directive, needs an asynchronous `with` context manager because of the connection's `AsyncEngine`, which needs to be closed after every `commit()` transaction. Closing the session factory is not part of the procedure in the synchronous ORM version.

- Second, after its creation, `AsyncSession` will only start executing all the CRUD transactions when the service calls its `begin()` method. The main reason is that `AsyncSession` can be closed and needs to be closed once the transaction has been executed. That is why another asynchronous context manager is used to manage `AsyncSession`.

The following code shows the `APIRouter` script, which implements the services for monitoring gym member attendance using the asynchronous `AttendanceRepository`:

```
from fastapi import APIRouter
from db_config.sqlalchemy_async_connect import
        AsynSessionFactory
from repository.sqlalchemy.attendance import
        AttendanceRepository
from models.requests.attendance import AttendanceMemberReq
from models.data.sqlalchemy_async_models import
        Attendance_Member


router = APIRouter()


@router.post("/attendance/add")
async def add_attendance(req:AttendanceMemberReq ):
    async with AsynSessionFactory() as sess:
        async with sess.begin():
            repo = AttendanceRepository(sess)
            attendance = Attendance_Member(id=req.id,
                member_id=req.member_id,
                timein=req.timein, timeout=req.timeout,
                date_log=req.date_log)
            return await repo.insert_attendance(attendance)
```

```
@router.patch("/attendance/update")
async def update_attendance(id:int,
                    req:AttendanceMemberReq ):
    async with AsynSessionFactory() as sess:
        async with sess.begin():
            repo = AttendanceRepository(sess)
            attendance_dict = req.dict(exclude_unset=True)
            return await repo.update_attendance(id,
                    attendance_dict)


@router.delete("/attendance/delete/{id}")
async def delete_attendance(id:int):
    async with AsynSessionFactory() as sess:
        async with sess.begin():
            repo = AttendanceRepository(sess)
            return await repo.delete_attencance(id)


@router.get("/attendance/list")
async def list_attendance():
    async with AsynSessionFactory() as sess:
        async with sess.begin():
            repo = AttendanceRepository(sess)
            return await repo.get_all_attendance()
```

The preceding script shows no direct parameter passing between the repository class and the AsyncSession instance. The session must comply with the two context managers before it becomes a working one. This syntax is valid under *SQLAlchemy 1.4*, which may undergo some changes in the future with SQLAlchemy's next releases.

Other ORM platforms that have been created for asynchronous transactions are easier to use. One of these is **GINO**.

Using GINO for async transactions

GINO, which stands for **GINO Is Not ORM**, is a lightweight asynchronous ORM that runs on top of an SQLAlchemy Core and AsyncIO environment. All its APIs are asynchronous-ready so that you can build contextual database connections and transactions. It has built-in *JSONB* support so that it can convert its results into JSON objects. But there is one catch: GINO only supports PostgreSQL databases.

While creating the gym fitness project, the only available stable GINO version is 1.0.1, which requires *SQLAlchemy 1.3*. Therefore, installing GINO will automatically uninstall *SQLAlchemy 1.4*, thus adding the GINO repository to the ch05a project to avoid any conflicts with the async version of SQLAlchemy.

You can use the following command to install the latest version of GINO:

```
pip install gino
```

Installing the database driver

Since the only RDBMS it supports is PostgreSQL, you only need to install `asyncpg` using the `pip` command.

Establishing the database connection

No other APIs are needed to open a connection to the database except for the `Gino` directive. We need to instantiate the class to start building the domain layer. The `Gino` class can be imported from the ORM's `gino` module, as shown by the following script:

```
from gino import Gino
db = Gino()
```

Its instance is like a façade that controls all database transactions. It starts by establishing a database connection once it's been provided with the correct PostgreSQL administrator credentials. The full GINO database connectivity script can be found in the `/db_config/gino_connect.py` script file. Let us now build the model layer.

Building the model layer

The model class definition in GINO has similarities with SQLAlchemy when it comes to structuring, column metadata, and even the existence of the `__tablename__` property. The only difference is the superclass type because GINO uses the `Model` class from the database reference instance's `db`. The following script shows how the `Signup` domain model is mapped to the `signup` table:

```
from db_config.gino_connect import db


class Signup(db.Model):
    __tablename__ = "signup"
    id = db.Column(db.Integer, primary_key=True,
                index=True)
    username = db.Column('username', db.String,
```

```
                    unique=False, index=False)
       password = db.Column('password',db.String,
                    unique=False, index=False)
```

Like in SQLAlchemy, the __tablename__ property is mandatory for all model classes to indicate their mapped table schema. When defining the column metadata, the db object has a Column directive that can set properties such as the *column type*, *primary key*, *unique*, *default*, *nullable*, and *index*. The column types also come from the db reference object, and these types are also the same for SQLAlchemy – that is, String, Integer, Date, Time, Unicode, and Float.

And just in case the name of the model attribute does not match the column name, the Column directive has its first parameter register the name of the actual column and maps it to the model attributes. The username and password columns are example cases of mapping the class attributes to the table's column names.

Mapping table relationships

At the time of writing, GINO only supports the *many-to-one relationship* by default. The db reference object has a ForeignKey directive, which establishes a foreign key relationship with the parent model. It just needs the actual reference key column and table name of the parent table to pursue the mapping. Setting the ForeignKey property in the Column object of the child model class is enough configuration to perform a *LEFT OUTER JOIN* to retrieve all the child records of the parent mode class. GINO has no relationship() function to address more details regarding how to fetch the child records of the parent model class. However, it has built-in loaders to automatically determine the foreign key and perform a many-to-one join query afterward. A perfect setup for this join query is the relationship configuration between the Profile_Trainers and Gym_Class model classes, as shown in the following script:

```
class Profile_Trainers(db.Model):
    __tablename__ = "profile_trainers"
    id = db.Column(db.Integer, db.ForeignKey('login.id'),
            primary_key=True, index=True)
    firstname = db.Column(db.String, unique=False,
            index=False)

... ... ... ... ... ...

    shift = db.Column(db.Integer, unique=False,
            index=False)


class Gym_Class(db.Model):
    __tablename__ = "gym_class"
    id = db.Column(db.Integer, primary_key=True,
```

```
                   index=True)
    member_id = db.Column(db.Integer,
        db.ForeignKey('profile_members.id'), unique=False,
           index=False)
    trainer_id = db.Column(db.Integer,
        db.ForeignKey('profile_trainers.id'), unique=False,
           index=False)
    approved = db.Column(db.Integer, unique=False,
       index=False)
```

We will have to make some changes if we need to build a query that will deal with a *one-to-many* or a *one-to-one relationship*. For the *LEFT OUTER JOIN* query to work, the parent model class must have a `set` collection defined to contain all the child records during join queries involving one-to-many relationships. For a *one-to-one relationship*, the parent only needs to instantiate the child model:

```
class Login(db.Model):
    __tablename__ = "login"
    id = db.Column(db.Integer, primary_key=True,
                index=True)
    username = db.Column(db.String, unique=False,
                index=False)
    … … … … … …
    def __init__(self, **kw):
        super().__init__(**kw)
        self._child = None

    @property
    def child(self):
        return self._child

    @child.setter
    def child(self, child):
        self._child = child

class Profile_Members(db.Model):
    __tablename__ = "profile_members"
    id = db.Column(db.Integer, db.ForeignKey('login.id'),
```

```
            primary_key=True, index=True)
... ... ... ... ... ...
weight = db.Column(db.Float, unique=False, index=False)
trainer_id = db.Column(db.Integer,
    db.ForeignKey('profile_trainers.id'), unique=False,
        index=False)

def __init__(self, **kw):
    super().__init__(**kw)
    self._children = set()


@property
def children(self):
    return self._children


@children.setter
def children(self, child):
    self._children.add(child)
```

This *set collection* or *child object* must be instantiated in the parent's __init__() to be accessed by the ORM's loader through the *children* or *child* @property, respectively. Using @property is the only way to manage joined records.

Note that the existence of the loader APIs is proof that GINO does not support the automated relationship that SQLAlchemy has. If we want to deviate from its core setup, Python programming is needed to add some features not supported by the platform, such as the one-to-many setup between Profile_Members and Gym_Class, and between Login and Profile_Members/Profile_Trainers. In the previous script, notice the inclusion of a constructor and the custom children Python property in Profile_Members, as well as the custom child property in Login. This is because GINO only has a built-in parent property.

You can find the domain models of GINO in the /models/data/gino_models.py script.

Important note

@property is a Python decorator that's used to implement a getter/setter in a class. This hides an instance variable from the accessor and exposes its *getter* and *setter* property *fields*. Using @property is one way to implement the *encapsulation* principle in Python.

Implementing the CRUD transactions

Let us consider the following `TrainerRepository`, which manages trainer profiles. Its `insert_trainer()` method shows the conventional way of implementing insert transactions. GINO requires its model class to call `create()`, an inherited method from the db reference object. All the column values are passed to the `create()` method through named parameters or as a bundle using `kwargs` before the record object is persisted. But GINO allows another insert option that uses the instance of the model derived by injecting column values into its constructor. The created instance has a method called `create()` that inserts the record object without requiring any parameters:

```python
from models.data.gino_models import Profile_Members,
        Profile_Trainers, Gym_Class
from datetime import date, time
from typing import List, Dict, Any


class TrainerRepository:

    async def insert_trainer(self,
            details:Dict[str, Any]) -> bool:
        try:
            await Profile_Trainers.create(**details)
        except Exception as e:
            print(e)
            return False
        return True
```

`update_trainer()` highlights how GINO updates table records. Based on the script, updating the table in the GINO way involves doing the following:

- First, it requires the `get()` class method of the model class to retrieve the record object with the `id` primary key.

- Second, the extracted record has an instance method called `update()` that will automatically modify the mapped row with the new data specified in its `kwargs` argument. The `apply()` method will commit the changes and close the transaction:

```python
    async def update_trainer(self, id:int,
            details:Dict[str, Any]) -> bool:
        try:
```

```
            trainer = await Profile_Trainers.get(id)
            await trainer.update(**details).apply()
        except:
            return False
    return True
```

Another option is to use the SQLAlchemy `ModelClass.update.values(ModelClass).where(expression)` clause, which, when applied to `update_trainer()`, will give us this final statement:

```
Profile_Trainers.update.values(**details).
    where(Profile_Trainers.id == id).gino.status()
```

Its `delete_trainer()` also follows the same approach as the GINO *update* transaction. This transaction is a two-step process, and the last step requires calling the `delete()` instance method of the extracted record object:

```
async def delete_trainer(self, id:int) -> bool:
    try:
        trainer = await Profile_Trainers.get(id)
        await trainer.delete()
    except:
        return False
    return True
```

On the other hand, `TrainerRepository` has two methods, `get_member()` and `get_all_member()`, which show how GINO constructs query statements:

- The former retrieves a specific record object using its primary key through the `get()` class method of the model class

- The latter uses the `gino` extension of `query` to utilize the `all()` method, which retrieves the records:

```
async def get_all_member(self):
    return await Profile_Trainers.query.gino.all()


async def get_member(self, id:int):
    return await Profile_Trainers.get(id)
```

But what translates database rows into model objects in a query's execution is the built-in loader of GINO. If we expand further on the solution presented in get_all_member(), this will look like this:

```
query = db.select([Profile_Trainers])
q = query.execution_options(
        loader=ModelLoader(Profile_Trainers))
users = await q.gino.all()
```

In the GINO ORM, all queries utilize ModelLoader to load each database record into a model object:

```
class GymClassRepository:

    async def join_classes_trainer(self):
        query = Gym_Class.join(Profile_Trainers).select()
        result = await query.gino.load(Gym_Class.
            distinct(Gym_Class.id).
                load(parent=Profile_Trainers)).all()
        return result

    async def join_member_classes(self):
        query = Gym_Class.join(Profile_Members).select()
        result = await query.gino.load(Profile_Members.
            distinct(Profile_Members.id).
                load(add_child=Gym_Class)).all()
        return result
```

If the normal query requires ModelLoader, what is needed for the *JOIN* query transactions? GINO has no automated support for table relationships, and creating *JOIN* queries is impossible without ModelLoader. The join_classes_trainer() method implements a *one-to-many* query for Profile_Trainers and Gym_Class. The distinct(Gym_Class.id). load(parent=Profile_Trainers) clause in the query creates a ModelLoader for GymClass, which will merge and load the Profile_Trainers parent record into its child Gym_Class. join_member_classes() creates *one-to-many* joins, while distinct(Profile_Members. id).load(add_child=Gym_Class) creates a ModelLoader to build the set of Gym_Class records, as per the Profile_Members parent.

On the other hand, the *many-to-one* relationship of Gym_Class and Profile_Members uses the load() function of Profile_Member, which is a different approach to matching the Gym_Class child records to Profile_Members. The following joined query is the opposite of the *one-to-many* setup because the Gym_Class records here are on the left-hand side while the profiles are on the right:

```
async def join_classes_member(self):
    result = await
        Profile_Members.load(add_child=Gym_Class)
        .query.gino.all()
```

So, the loader plays an important role in building queries in GINO, especially joins. Although it makes query building difficult, it still gives flexibility to many complex queries.

All the repository classes for GINO can be found in the /repository/gino/trainers.py script.

Running the CRUD transactions

For our repositories to run in the APIRouter module, we need to open the database connection by binding the db reference object to the actual database through DB_URL. It is ideal to use a dependable function for the binding procedure because the easier form of rolling out is done through APIRouter injection. The following script shows how to set up this database binding:

```
from fastapi import APIRouter, Depends
from fastapi.encoders import jsonable_encoder
from fastapi.responses import JSONResponse
from db_config.gino_connect import db
from models.requests.trainers import ProfileTrainersReq
from repository.gino.trainers import TrainerRepository


async def sess_db():
    await db.set_bind(
      "postgresql+asyncpg://
        postgres:admin2255@localhost:5433/fcms")


router = APIRouter(dependencies=[Depends(sess_db)])


@router.patch("/trainer/update" )
async def update_trainer(id:int, req: ProfileTrainersReq):
    mem_profile_dict = req.dict(exclude_unset=True)
    repo = TrainerRepository()
```

```
    result = await repo.update_trainer(id,
            mem_profile_dict)
    if result == True:
        return req
    else:
        return JSONResponse(
    content={'message':'update trainer profile problem
            encountered'}, status_code=500)


@router.get("/trainer/list")
async def list_trainers():
    repo = TrainerRepository()
    return await repo.get_all_member()
```

The list_trainers() and update_trainer() REST services shown in the preceding code
are some services of our *fitness club* application that will successfully run TrainerRepository
after injecting sess_db() into APIRouter. GINO does not ask for many details when establishing
the connection to PostgreSQL except for DB_URL. Always specify the asyncpg dialect in the URL
because it is the only driver that's supported by GINO as a synchronous ORM.

Creating the tables

GINO and SQLAlchemy have the same approach to creating a table schema at the framework level.
Both require the MetaData and Column directives for building the Table definitions. Then, an
asynchronous function is preferred to derive the engine using the create_engine() method with
our DB_URL. Like in SQLAlchemy, this engine plays a crucial role in building the tables through
create_all(), but this time, it uses GINO's GinoSchemaVisitor instance. The following script
shows the complete implementation of how tables are generated in GINO using the AsyncIO platform:

```
from sqlalchemy import Table, Column, Integer, String,
            MetaData, ForeignKey'
import gino
from gino.schema import GinoSchemaVisitor


metadata = MetaData()


signup = Table(
    'signup', metadata,
```

```
        Column('id', Integer, primary_key=True),
        Column('username', String),
        Column('password', String),
    )

    ... ... ... ... ...
async def db_create_tbl():
    engine = await gino.create_engine(DB_URL)
    await GinoSchemaVisitor(metadata).create_all(engine)
```

As stated in SQLAlchemy, performing DDL transactions such as schema auto-generation at the start is optional because it may cause FastAPI's performance to degrade, and even some conflicts in the existing database schema.

Now, let us explore another ORM that requires custom Python coding: **Pony ORM**.

Using Pony ORM for the repository layer

Pony ORM relies on Python syntax for building the model classes and repository transactions. This ORM only uses Python data types such as int, str, and float, as well as class types to implement model definitions. It applies Python lambda expressions to establish CRUD transactions, especially when mapping table relationships. Also, Pony heavily supports JSON conversion of record objects when reading records. On the other hand, Pony can cache the query objects, which provides faster performance than the others. The code for Pony ORM can be found in the ch05a project.

To use Pony, we need to install it using pip. This is because it is a third-party platform:

```
pip install pony
```

Installing the database driver

Since Pony is an ORM designed to build synchronous transactions, we will need the psycopg2 PostgreSQL driver. We can install it using the pip command:

```
pip install psycopg2
```

Creating the database's connectivity

The approach to establishing database connectivity in Pony is simple and declarative. It only needs the `Database` directive from the `pony.orm` module to be instantiated to connect to the database using the correct database credentials. The following script is used in the *fitness club* prototype:

```
from pony.orm import  Database

db = Database("postgres", host="localhost", port="5433",
    user="postgres", password="admin2255", database="fcms")
```

As you can see, the first parameter of the constructor is the *database dialect*, followed by kwargs, which contains all the details about the connection. The full configuration can be found in the `/db_config/pony_connect.py` script file. Now, let us create Pony's model classes.

Defining the model classes

The created database object, db, is the only component needed to define a Pony *entity*, a term that refers to a model class. It has an `Entity` attribute, which is used to subclass each model class to provide the `_table_` attribute, which is responsible for the *table-entity* mapping. All entity instances are bound to db and mapped to the tables. The following script shows how the `Signup` class becomes an entity of the model layer:

```
from pony.orm import  Database, PrimaryKey, Required,
        Optional, Set
from db_config.pony_connect import db
from datetime import date, time


class Signup(db.Entity):
    _table_ = "signup"
    id = PrimaryKey(int)
    username = Required(str, unique=True, max_len=100,
        nullable=False, column='username')
    password = Required(str, unique=Fals, max_len=100,
        nullable=False, column='password')
```

The pony.orm module contains Required, Optional, PrimaryKey, or Set directives, which are used to create column attributes. Since each entity must have a primary key, PrimaryKey is used to define the column attribute of the entity. If the class has no primary key, Pony ORM will implicitly generate an id primary for the entity with the following definition:

```
id = PrimaryKey(int, auto=True)
```

On the other hand, the Set directive indicates relationships between entities. All these directives have a mandatory attribute column type, which declares the column value type in Python syntax (for example, int, str, float, date, or time) or any class type. Other column attributes include auto, max_len, index, unique, nullable, default, and column. Now, let us establish a relationship between model classes:

```
class Login(db.Entity):
    _table_ = "login"
    id = PrimaryKey(int)

    ... ... ... ... ... ...

    date_approved = Required(date)
    user_type = Required(int)

    trainers = Optional("Profile_Trainers", reverse="id")
    members = Optional("Profile_Members", reverse="id")
```

The given Login class has two additional attributes, trainers and members, which serve as reference keys to the Profile_Trainers and Profile_Members models, respectively. In turn, these child entities have their respective class attributes pointing back at the Login model, establishing a relationship. These column attributes and their reference-foreign keys relationship must match the physical database schema. The following code shows examples of Pony's child model classes:

```
class Profile_Trainers(db.Entity):
    _table_ = "profile_trainers"
    id = PrimaryKey("Login", reverse="trainers")
    firstname = Required(str)

    ... ... ... ... ... ...

    tenure = Required(float)
    shift = Required(int)

    members = Set("Profile_Members",
            reverse="trainer_id")
    gclass = Set("Gym_Class", reverse="trainer_id")
```

```
class Profile_Members(db.Entity):
    _table_ = "profile_members"
    id = PrimaryKey("Login", reverse="members")
    firstname = Required(str)
    ... ... ... ... ... ...
    trainer_id = Required("Profile_Trainers",
            reverse="members")
    ... ... ... ... ... ...
```

Defining the relationship attributes depends on the relationship type between the two entities. Attributes should be defined as *Optional(parent)-Required(child)* or *Optional(parent)-Optional(child)* if the relationship type is one-to-one. For one-to-many, attributes should be defined as *Set(parent)-Required(child)*. Finally, for many-to-one, the attributes must be defined as *Set(parent)-Set(child)*.

`Login` has a one-to-one relationship with `Profile_Members`, which explains the use of the `Optional` attribute to point to the `id` key of `Profile_Members`. The primary keys are always the reference keys in this relationship for Pony.

On the other hand, the `Profile_Trainers` model has a one-to-many setup with `Profile_Members`, which explains why the `trainer_id` attribute of the former uses the `Required` directive to point to the `Set` attribute `members` of the latter. Sometimes, the framework requires backreference through the directive's `reverse` parameter. The preceding code also depicts the same scenario between the `Profile_Members` and `Gym_Class` models, where the `gclass` attribute of `Profile_Members` is declared as a `Set` collection that contains all the enrolled gym classes of the member. The reference key can be a primary key or just a typical class attribute in this relationship. The following snippet shows the blueprint of the `Gym_Class` model:

```
class Gym_Class(db.Entity):
    _table_ = "gym_class"
    id = PrimaryKey(int)
    member_id = Required("Profile_Members",
        reverse="gclass")
    trainer_id = Required("Profile_Trainers",
        reverse="gclass")
    approved = Required(int)
db.generate_mapping()
```

Unlike in other ORMs, Pony needs `generate_mapping()` to be executed to realize all the entity mappings to the actual tables. The method is an instance method of the `db` instance that must appear in the last part of the module script, as shown in the previous snippet, where `Gym_Class` was the last Pony model class to be defined. All the Pony model classes can be found in the `/models/data/pony_models.py` script file.

Note that we can create Pony entities manually or digitally using *Pony ORM ER Diagram Editor*, which we can access at `https://editor.ponyorm.com/`. The editor can provide us with both free and commercial accounts. Let us now implement the CRUD transactions.

Implementing the CRUD transactions

CRUD transactions in Pony are session-driven. But unlike SQLAlchemy, its repository classes do not require injecting `db_session` into the repository constructor. Each transaction in Pony will not work without `db_session`. The following code shows a repository class that implements all the transactions needed to manage a list of gym members:

```python
from pony.orm import db_session, left_join
from models.data.pony_models import Profile_Members,
            Gym_Class, Profile_Trainers
from datetime import date, time
from typing import List, Dict, Any
from models.requests.members import ProfileMembersReq


class MemberRepository:


    def insert_member(self,
            details:Dict[str, Any]) -> bool:
        try:
            with db_session:
                Profile_Members(**details)
        except:
            return False
        return True
```

In Pony, inserting a record means instantiating the model class with the injected record values. An example is `insert_member()`, which inserts a profile by instantiating the `Profile_Members` model with the injected membership details. However, the case is different when updating records, as shown in the following script:

```
def update_member(self, id:int,
          details:Dict[str, Any]) -> bool:
    try:
        with db_session:
            profile = Profile_Members[id]
            profile.id = details["id"]
            ... ... ... ... ... ...
            profile.trainer_id = details["trainer_id"]
    except:
        return False
    return True
```

Updating a record in Pony, which is implemented in the `update_member()` script, means retrieving the record object through *indexing* using its `id`. The retrieved object is automatically converted into a JSON-able object since Pony has built-in support for JSON. Then, the new values of those attributes are overwritten as they must be changed. This *UPDATE* transaction is, again, within the bounds of `db_session`, thus automatically refreshing the record(s) after the overwrites.

On the other hand, `delete_member()` of the repository class shows the same approach with *UPDATE*, except that a `delete()` class method is invoked right after retrieving the object record. The following is the script for this operation:

```
def delete_member(self, id:int) -> bool:
    try:
        with db_session:
            Profile_Members[id].delete()
    except:
        return False
    return True
```

The delete transaction is also `db_session` bound, so invoking `delete()` automatically refreshes the table. The following code shows Pony's implementation for query transactions:

```
def get_all_member(self):
    with db_session:
```

```
        members = Profile_Members.select()
        result = [ProfileMembersReq.from_orm(m)
            for m in members]
        return result

    def get_member(self, id:int):
        with db_session:
            login = Login.get(lambda l: l.id == id)
            member = Profile_Members.get(
                lambda m: m.id == login)
            result = ProfileMembersReq.from_orm(member)
        return result
```

get_member() retrieves a single record using the get() class method, which requires a lambda expression in its parameter. Since Login has a one-to-one relationship with Profile_Members, first, we must extract the Login record of the member and use the login object to fetch the record through the get() helper function of the Profile_Members entity. This approach is also applicable to other entities with other entity relationships. Now, get_all_member() retrieves a result set using the select() method. The select() method can also utilize a lambda expression if there are constraints in the retrieval operation.

Pony model classes have the get() and select() methods, which both return Query objects that FastAPI cannot process directly. So, we need an ORM-friendly Pydantic model to extract the final entities from these Query objects. Like in SQLAlchemy, a ModelBase class with a nested Config class is required to retrieve the records from the Query object. The nested class must configure orm_mode to True. If relationship mappings are involved, the request model must also declare the attributes involved in the relationship and their corresponding child object converters. The method converters, decorated by Pydantic's @validator, are automatically called by Pony to interpret and validate the Query objects into JSON-able components such as List or entity objects. The following code shows the request model that's used to extract the records from select() through list comprehension and the Profile_Member dict object from get():

```
from typing import List, Any
from pydantic import BaseModel, validator


class ProfileMembersReq(BaseModel):
    id: Any
    firstname: str
    lastname: str
    age: int
```

```
    height: float
    weight: float
    membership_type: str
    trainer_id: Any

    gclass: List

    @validator('gclass', pre=True,
        allow_reuse=True, check_fields=False)
    def gclass_set_to_list(cls, values):
        return [v.to_dict() for v in values]

    @validator('trainer_id', pre=True,
        allow_reuse=True, check_fields=False)
    def trainer_object_to_map(cls, values):
        return values.to_dict()

    class Config:
        orm_mode = True
```

The presence of the `gclass_set_to_list ()` and `trainer_object_to_map()` converts in `ProfileMembersReq` enables data to be populated to the child objects in the `gclass` and `trainer_id` attributes, respectively. These additional features indicate why executing `select()` can already retrieve the *INNER JOIN* queries.

To build *LEFT JOIN* query transactions, the ORM has a built-in directive called `left_join()`, which is used to extract the `Query` object bearing the *LEFT JOIN* raw objects through a Python generator. The following code shows another repository class that showcases the use of `left_join()`:

```
class MemberGymClassRepository:

    def join_member_class(self):
      with db_session:
        generator_args = (m for m in Profile_Members
                for g in m.gclass)
        joins = left_join(tuple_args)
        result = [ProfileMembersReq.from_orm(m)
```

```
                for m in joins ]
        return result
```

All the repository classes can be found in the `/repository/pony/members.py` script file.

Now, what makes Pony faster is that it uses an *identity map*, which contains all the record objects that have been retrieved from every single query transaction. The ORM applies the *Identity Map* design pattern to apply its caching mechanism to make read and write executions fast. It only requires memory management and monitoring to avoid memory leak problems in complex and huge applications.

Running the repository transactions

Since `db_session` is already managed internally, no additional requirements will be needed from Pony for the `APIRouter` script to run the repository transactions. The repository classes are directly accessed and instantiated in each of the APIs to access the CRUD transactions.

Creating the tables

If the tables are non-existent yet, Pony can generate those tables through its entity classes. This DDL transaction is enabled when the `create_tables` parameter of the `generate_mapping()` method of `db` is set to `True`.

For the most compact and simplest ORM in terms of syntax, we have **Peewee**.

Building the repository using Peewee

Among the different ORMs, Peewee is the simplest and smallest in terms of ORM features and APIs. The framework is easy to understand and use; it is not comprehensive, but it has intuitive ORM syntax. Its strength is in building and executing query transactions.

Peewee is not designed for asynchronous platforms, but it can work with them by using some async-related libraries that it supports. We need to install at least *Python 3.7* for Peewee to work with FastAPI, an asynchronous framework. To install Peewee, we need to execute the following command:

```
pip install peewee
```

Installing the database driver

The ORM needs `psycopg2` as the PostgreSQL database driver. We can install it using `pip`:

```
pip install psycopg2
```

Creating the database connection

For Peewee to work with FastAPI, we must build a multi-threading mechanism where Peewee can cater to more than one request transaction on the same thread, and per request can do more executions simultaneously using different local threads. This customized multi-threading component, which can be created using the `ContextVar` class, bridges Peewee to the FastAPI platform. But for Peewee to utilize these threads, we also need to customize its `_ConnectionState` with the newly created threading state, `db_state`. The following code shows how `db_state` and a custom `_ConnectionState` can be derived:

```
from peewee import _ConnectionState
from contextvars import ContextVar


db_state_default = {"closed": None, "conn": None,
        "ctx": None, "transactions": None}
db_state = ContextVar("db_state",
        default=db_state_default.copy())


class PeeweeConnectionState(_ConnectionState):
    def __init__(self, **kwargs):
        super().__setattr__("_state", db_state)
        super().__init__(**kwargs)

    def __setattr__(self, name, value):
        self._state.get()[name] = value

    def __getattr__(self, name):
        return self._state.get()[name]
```

To apply the new `db_state` and `_ConnectionState` classes, cited in the preceding code as `PeeweeConnectionState`, we need to open the database connection through the `Database` class. Peewee has several variations of the `Database` class, depending on the type of database the application will choose to connect to. Since we will be using PostgreSQL, `PostgresqlDatabase` is the correct class to initialize with all the necessary database details. After establishing the connection, the `db` instance will have a `_state` attribute that will point to the `PeeweeConnectionState` instance. The following snippet shows how to connect to our *fitness gym* database's `fcms` using the database credentials:

```
from peewee import PostgresqlDatabase
```

```
db = PostgresqlDatabase(
    'fcms',
    user='postgres',
    password='admin2255',
    host='localhost',
    port=5433,
)


db._state = PeeweeConnectionState()
```

The preceding code also emphasizes that the default state of the database connection must be replaced with a non-blocking one that can work with the FastAPI platform. This configuration can be found in the /db_config/peewee_connect.py script file. Let us now build Peewee's model layer.

Creating the tables and the domain layer

Peewee prefers *auto-generating tables* based on its model classes, unlike other ORMs. Peewee recommends *reverse engineering*, where tables are created rather than only being mapped to existing tables. Letting the application generate the tables lessens the hassle of establishing relationships and primary keys. This ORM is unique because it has an "implied" approach to creating primary keys and foreign keys. The following script shows how Peewee model classes are defined:

```
from peewee import Model, ForeignKeyField, CharField,
    IntegerField, FloatField, DateField, TimeField
from db_config.peewee_connect import db


class Signup(Model):
    username = CharField(unique=False, index=False)
    password = CharField(unique=False, index=False)

    class Meta:
        database = db
        db_table = 'signup'
```

We can't see any primary keys in the model classes presented because the Peewee engine will create them during its schema auto-generation. The physical foreign key column and the model attribute will have the same name derived from its model name, with the `modelname_id` pattern in lowercase form. If we insist on adding the primary key for the model, a conflict will occur, making Peewee dysfunctional. We must let Peewee create the physical tables from the model classes to avoid this mishap.

All model classes inherit properties from the `Model` directive of the ORM. It also has column directives such as `IntegerField`, `FloatField`, `DateField`, and `TimeField` for defining the column attributes of the model classes. Moreover, each domain class has a nested `Meta` class, which registers the references to `database` and `db_table`, which is mapped to the model class. Other properties that we can set here are `primary_key`, `indexes`, and `constraints`.

The only problem in having this auto-generation is when creating table relationships. Linking the foreign key attributes of the child classes to the non-existent primary keys of the parent classes is difficult before auto-generation. For instance, the following `Profile_Trainers` model implies a *many-to-one* relationship with the `Login` class, which is only defined by the `ForeignKeyField` directive with the `trainer` backreference attribute and not by the `login_id` foreign key:

```
class Profile_Trainers(Model):
    login = ForeignKeyField(Login,
        backref="trainers", unique=True)
    ... ... ... ... ... ...
    shift = IntegerField(unique=False, index=False)

    class Meta:
      database = db
      db_table = 'profile_trainers'
```

The `login_id` column that's generated after auto-generation can be seen in the following screenshot:

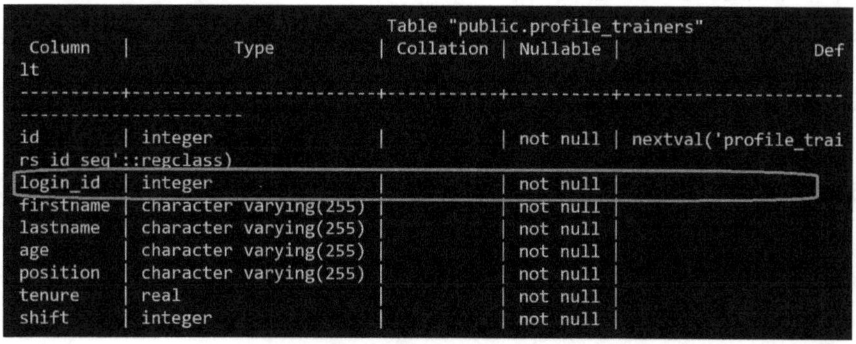

```
                                Table "public.profile_trainers"
   Column    |           Type            | Collation | Nullable |                   Def
   lt
-------------+---------------------------+-----------+----------+-------------------------
---------------------------
 id          | integer                   |           | not null | nextval('profile_trai
rs_id_seq'::regclass)
 login_id    | integer                   |           | not null |
 firstname   | character varying(255)    |           | not null |
 lastname    | character varying(255)    |           | not null |
 age         | character varying(255)    |           | not null |
 position    | character varying(255)    |           | not null |
 tenure      | real                      |           | not null |
 shift       | integer                   |           | not null |
```

Figure 5.2 – The generated profile_trainers schema

Foreign key attributes are declared using the `ForeignKeyField` directive, which accepts at least three crucial parameters:

- The parent model's name
- The `backref` parameter, which references the child record (if in a *one-to-one* relationship) or a set of child objects (if in a *one-to-many* or *many-to-one* relationship)
- The `unique` parameter, which indicates a *one-to-one* relationship when set to `True` or `False` otherwise

After defining all the models, including their relationships, we need to call the following methods from Peewee's db instance for the table mapping to occur:

- `connect()` to establish the connection
- `create_tables()` to pursue the schema generation based on its list of model classes

The following script shows a snapshot of the class definitions, including the call to the two db methods:

```python
class Login(Model):
    username = CharField(unique=False, index=False)
    ... ... ... ... ... ...
    user_type = IntegerField(unique=False, index=False)

    class Meta:
        database = db
        db_table = 'login'


class Gym_Class(Model):
    member = ForeignKeyField(Profile_Members,
            backref="members")
    trainer = ForeignKeyField(Profile_Trainers,
            backref="trainers")
    approved = IntegerField(unique=False, index=False)

    class Meta:
        database = db
        db_table = 'gym_class'


db.connect()
```

```
db.create_tables([Signup, Login, Profile_Members,
    Profile_Trainers, Attendance_Member, Gym_Class],
        safe=True)
```

As we can see, we need to set the `safe` parameter of `create_tables()` to True so that Peewee will only perform schema auto-generation once during the initial server startup of the application. All the model classes for the Peewee ORM can be found in the `/models/data/peewee_models.py` script file. Now, let us implement the repository layer.

Implementing the CRUD transactions

Creating the asynchronous connection and building the model layer for the application in the Peewee ORM is tricky, but implementing its repository layer is straightforward. All the method operations are entirely derived from its model classes. For instance, `insert_login()`, which is shown in the following snippet, shows how the `create()` static method of `Login` takes the login details for record insertion:

```
from typing import Dict, List, Any
from models.data.peewee_models import Login,
    Profile_Trainers, Gym_Class, Profile_Members
from datetime import date


class LoginRepository:

    def insert_login(self, id:int, user:str, passwd:str,
            approved:date, type:int) -> bool:
        try:
            Login.create(id=id, username=user,
                password=passwd, date_approved=approved,
                user_type=type)
        except Exception as e:
            return False
        return True
```

This method can be re-implemented to perform bulk inserts, but Peewee has an alternative way to pursue multiple insertions through its `insert_many()` class method. Using `insert_many()` requires more accurate column details for mapping multiple schema values. It also needs an invocation of the `execute()` method to perform all the bulk insertions and close the operation afterward.

Similarly, the update() class method requires the execute() method after filtering the record that needs updating using the id primary key. This is shown in the following code snippet:

```
def update_login(self, id:int,
        details:Dict[str, Any]) -> bool:
    try:
        query = Login.update(**details).
            where(Login.id == id)
        query.execute()
    except:
        return False
    return True
```

When it comes to record deletion, delete_login() shows the easy approach – that is, by using delete_by_id(). But the ORM has another way, which is to retrieve the record object using the get() class method – for example, Login.get(Login.id == id) – and eventually delete the record through the delete_instance() instance method of the record object. The following delete_login() transaction shows how to utilize the delete_by_id() class method:

```
def delete_login(self, id:int) -> bool:
    try:
        query = Login.delete_by_id(id)
    except:
        return False
    return True
```

The following scripts, which are for get_all_login() and get_login(), highlight how Peewee retrieves records from the database. Peewee uses its get() class method to retrieve a single record using the primary key; the same method was applied to its *UPDATE* transaction in the previous code snippet. Similarly, Peewee uses a class method to extract multiple records, but this time, it uses the select() method. The resulting object can't be encoded by FastAPI unless it's contained in the List collection, which serializes the rows of data into a list of JSON-able objects:

```
def get_all_login(self):
    return list(Login.select())


def get_login(self, id:int):
    return Login.get(Login.id == id)
```

On the other hand, the following repository classes show how to create *JOIN* queries using its `join()` method:

```
from peewee import JOIN

class LoginTrainersRepository:

    def join_login_trainers(self):
        return list(Profile_Trainers.
            select(Profile_Trainers, Login).join(Login))

class MemberGymClassesRepository:
    def outer_join_member_gym(self):
        return list(Profile_Members.
            select(Profile_Members,Gym_Class).join(Gym_Class,
                join_type=JOIN.LEFT_OUTER))
```

`join_login_trainers()` of `LoginTrainersRepository` builds the *INNER JOIN* query of the `Profile_Trainers` and `Login` objects. The leftmost model indicated in the parameter of the `Profile_Trainers` object's `select()` directive is the parent model type, followed by its child model class in a *one-to-one* relationship. The `select()` directive emits the `join()` method with the model class type, which indicates the type of records that belong to the right-hand side of the query. The *ON* condition(s) and the foreign key constraints are optional but can be declared explicitly by adding the `on` and `join_type` attributes of the `join()` construct. An example of this query is `outer_join_member_gym()` of `MemberGymClassesRepository`, which implements a *LEFT OUTER JOIN* of `Profile_Members` and `Gym_Class` using the `LEFT_OUTER` option of the `join_type` attribute of `join()`.

Joins in Peewee also need the `list()` collection to serialize the retrieved records. All the repository classes for Peewee can be found in the `/repository/peewee/login.py` script.

Running the CRUD transaction

Since Peewee's database connection is set at the model layer, no additional requirements are required for `APIRouter` or `FastAPI` to run the CRUD transactions. API services can easily access all the repository classes without calling methods or directives from the db instance.

So far, we have experimented with popular ORMs to integrate a relational database into the FastAPI framework. If applying an ORM is not enough for a microservice architecture, we can utilize some design patterns that can further refine the CRUD performance, such as **CQRS**.

Applying the CQRS design pattern

CQRS is a microservice design pattern responsible for segregating query transactions (*reads*) from the insert, update, and delete operations (*writes*). The separation of these two groups lessens the cohesion access to these transactions, which provides less traffic and faster performance, especially when the application becomes complex. Moreover, this design pattern creates a loose-coupling feature between the API services and the repository layer, which gives us an advantage if there are several turnovers and changes in the repository layers.

Defining the handler interfaces

To pursue CQRS, we need to create the two interfaces that define the query and the command transactions. The following code shows the interfaces that will identify the *read* and *write* transactions for `Profile_Trainers`:

```
class IQueryHandler:
    pass


class ICommandHandler:
    pass
```

Here, `IQueryHandler` and `ICommandHandler` are informal interfaces because Python does not have an actual definition of an interface.

Creating the command and query classes

Next, we need to implement the command and query classes. The command serves as an instruction to pursue the *write* transactions. It also carries the state of the result after they have been executed. On the other hand, the query instructs the *read* transaction to retrieve record(s) from the database and contain the result afterward. Both components are serializable classes with *getter/setter* attributes. The following code shows the script for `ProfileTrainerCommand`, which uses Python's `@property` attribute to store the state of the *INSERT* execution:

```
from typing import Dict, Any


class ProfileTrainerCommand:

    def __init__(self):
        self._details:Dict[str,Any] = dict()

    @property
```

```
    def details(self):
        return self._details

    @details.setter
    def details(self, details):
        self._details = details
```

The details property will store all the column values of the trainer's profile record that need to be persisted.

The following script implements a sample *query* class:

```
class ProfileTrainerListQuery:

    def __init__(self):
        self._records:List[Profile_Trainers] = list()

    @property
    def records(self):
        return self._records

    @records.setter
    def records(self, records):
        self._records = records
```

The constructor of ProfileTrainerListQuery prepares a dictionary object that will contain all the retrieved records after the query transaction has been executed.

Creating the command and query handlers

We will be using our previous interfaces to define the command and query handlers. Note that the command handler accesses and executes the repository to execute the *write* transactions, while the query handler processes the *read* transactions. These handlers serve as the façade between the API services and the repository layer. The following code shows the script for AddTrainerCommandHandler, which manages the *INSERT* transaction for the trainer's profile:

```
from cqrs.handlers import ICommandHandler
from repository.gino.trainers import TrainerRepository
from cqrs.commands import ProfileTrainerCommand
```

```python
class AddTrainerCommandHandler(ICommandHandler):

    def __init__(self):
        self.repo:TrainerRepository = TrainerRepository()

    async def handle(self,
            command:ProfileTrainerCommand) -> bool:
        result = await self.repo.
            insert_trainer(command.details)
        return result
```

The handler depends on `ProfileTrainerCommand` for the record values that are crucial to the asynchronous execution of its `handle()` method.

The following script shows a sample implementation for a query handler:

```python
class ListTrainerQueryHandler(IQueryHandler):
    def __init__(self):
        self.repo:TrainerRepository = TrainerRepository()
        self.query:ProfileTrainerListQuery =
            ProfileTrainerListQuery()

    async def handle(self) -> ProfileTrainerListQuery:
        data = await self.repo.get_all_member();
        self.query.records = data
        return self.query
```

Query handlers return their *query* to the services and not the actual values. The `handle()` method of `ListTrainerQueryHandler` returns `ProfileTrainerListQuery`, which contains the list of records from the *read* transaction. This mechanism is one of the main objectives of applying CQRS to microservices.

Accessing the handlers

CQRS, aside from managing the friction between the *read* and *write* executions, does not allow the API services to interact directly with the execution of CRUD transactions. Moreover, it streamlines and simplifies the access of the CRUD transactions by assigning only the handler that's needed by a particular service.

The following script shows how AddTrainerCommand is only directly associated with the add_ trainer() service and how LisTrainerQueryHandler is only directly associated with the list_trainers() service:

```
from cqrs.commands import ProfileTrainerCommand
from cqrs.queries import ProfileTrainerListQuery
from cqrs.trainers.command.create_handlers import
    AddTrainerCommandHandler
from cqrs.trainers.query.query_handlers import
    ListTrainerQueryHandler

router = APIRouter(dependencies=[Depends(get_db)])


@router.post("/trainer/add" )
async def add_trainer(req: ProfileTrainersReq):
    handler = AddTrainerCommandHandler()
    mem_profile = dict()
    mem_profile["id"] = req.id
    … … … … … …
    mem_profile["shift"] = req.shift
    command = ProfileTrainerCommand()
    command.details = mem_profile
    result = await handler.handle(command)
    if result == True:
        return req
    else:
        return JSONResponse(content={'message':'create
          trainer profile problem encountered'},
            status_code=500)


@router.get("/trainer/list")
async def list_trainers():
    handler = ListTrainerQueryHandler()
    query:ProfileTrainerListQuery = await handler.handle()
    return query.records
```

We can identify transactions that are accessed frequently in `APIRouter` through CQRS. It helps us find which transactions need performance tuning and focus, which can help us avoid performance issues when the amount of access increases. When it comes to enhancement and upgrades, the design pattern can help developers find what domain to prioritize because of the separation of aspects in the repository layer. Generally, it offers flexibility to the application when its business processes need to be revamped. All the CQRS-related scripts can be found in the `/cqrs/` project folder.

Summary

Applying ORM always has advantages and disadvantages for any application. It can bloat the application with so many configurations and layers of components, and it can even slow down the application if not managed well. But ORM, in general, can help optimize query development by simplifying the constructs by using its APIs and eliminating unimportant repetitive SQL scripts. Overall, it can reduce the time and cost of software development compared to using `cursor` from `psycopg2`.

In this chapter, four Python ORMs were used, studied, and experimented with to help FastAPI create its repository layer. First, there is *SQLAlchemy*, which provides a boilerplated approach to creating standard and asynchronous data persistency and query operations. Then, there is *GINO*, which uses the AsyncIO environment to implement asynchronous CRUD transactions with its handy syntax. Also, there is *Pony*, the most Pythonic among the ORMs presented because it uses hardcore Python code to build its repository transactions. Lastly, there is *Peewee*, known for its concise syntax but tricky composition for the asynchronous database connection and CRUD transactions. Each ORM has its strengths and weaknesses, but all provide a logical solution rather than applying brute-force and native SQL.

If the ORM needs fine-tuning, we can add some degree of optimization by using data-related design patterns such as CQRS, which minimizes friction between the "read" and "write" CRUD transactions.

This chapter has highlighted the flexibility of FastAPI when utilizing ORMs to establish a connection to a relational database such as PostgreSQL. But what if we use a NoSQL database such as MongoDB to store information? Will FastAPI perform with the same level of performance when performing CRUD to and from MongoDB? The next chapter will discuss various solutions for integrating FastAPI into MongoDB.

Using a Non-Relational Database

So far, we have learned that relational databases store data using table columns and rows. All these table records are structurally optimized and designed using different keys, such as primary, unique, and composite keys. The tables are connected using foreign/reference keys. Foreign key integrity plays a significant role in the table relationship of a database schema because it gives consistency and integrity to the data that's persisted in the tables. *Chapter 5, Connecting to a Relational Database*, provided considerable proof that FastAPI can connect to relational databases using any of the present ORMs smoothly without lots of complexities. This time, we will focus on using non-relational databases as data storage for our FastAPI microservice application.

If FastAPI uses ORM for relational databases, it uses **Object Document Mapping** (**ODM**) to manage data using non-relational data stores or **NoSQL** databases. There are no tables, keys, and foreign key constraints involved in ODM, but a JSON document is needed to hold the various pieces of information. Different NoSQL databases vary in the storage model type that's used to store data. The simplest among these databases manages data as key-value pairs, such as **Redis**, while complicated databases utilize schema-free document structures easily mapped to objects. This is usually done in **MongoDB**. Some use columnar data stores such as **Cassandra**, while some have graph-oriented data storage such as **Neo4j**. However, this chapter will focus on the FastAPI-MongoDB connectivity and the different ODM we can apply to pursue data management with a document-based database.

The main objective of this chapter is to study, formalize, and scrutinize different ways to use MongoDB as a database for our FastAPI application. Building the repository layer and showcasing the CRUD implementation will be the main highlight.

In this chapter, we will cover the following topics:

- Setting up the database environment
- Applying the PyMongo driver for synchronous connections

- Creating asynchronous CRUD transactions using Motor

- Implementing CRUD transactions using MongoEngine

- Implementing asynchronous CRUD transactions using Beanie

- Building an asynchronous repository for FastAPI using ODMantic

- Creating CRUD transactions using MongoFrames

Technical requirements

This chapter focuses on an eBookstore web portal, *online book reselling system*, where users can sell and buy books from home through the internet. The virtual store allows users to view the *sellers' profiles*, *book catalogs*, *list of orders*, and *archive of purchases*. When it comes to the e-commerce side, the user can select their preferred books and add them to a cart. Then, they can check out the items as orders and pursue the payment transaction afterward. All the data is stored in a MongoDB database. The code for this chapter can be found at `https://github.com/PacktPublishing/Building-Python-Microservices-with-FastAPI` in the `ch06` project.

Setting up the database environment

Before we start discussing the application's database connectivity, we need to download the appropriate MongoDB database server from `https://www.mongodb.com/try/download/community`. *online book reselling system* uses MongoDB 5.0.5 for a Windows platform. The installation will provide default service configuration details for the service name, data directory, and log directory. However, it is advised that you use different directory paths instead of the default ones.

After the installation, we can start the MongoDB server by running `/bin/mongod.exe`. This will automatically create a database directory called `/data/db` in the `C:/` drive (Windows). We can place the `/data/db` directory in some other location, but be sure to run the `mongod` command with the `--dbpath` option while specifying `<new path>/data/db`.

The MongoDB platform has utilities that can aid in managing database collections, and one of them is **MongoDB Compass**. It can provide a GUI experience that allows you to browse, explore, and easily manipulate the database and its collections. Also, it has built-in performance metrics, query views, and schema visualization features that can help with scrutinizing the correctness of the database structure. The following screenshot shows the dashboard for MongoDB Compass version 1.29.6:

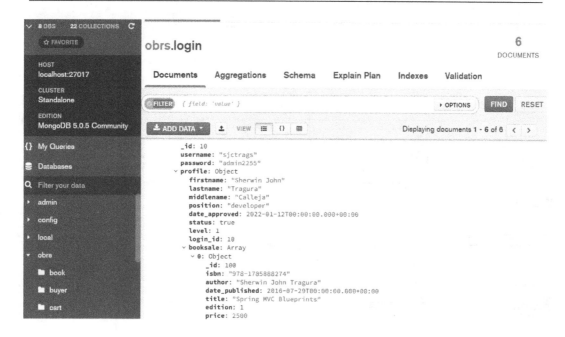

Figure 6.1 – The MongoDB Compass dashboard

The preceding dashboard shows the document structure of the **login** collection, which is part of the **obrs** database. It gives us the spread of the data, which is an easy way to view its embedded documents, such as `profile` and list of books for sale.

Once the server and utility have been installed, we need to design the data collections for our database using the **class diagram**. A class diagram is a UML approach to describing the components of a class and visualizing the associations and structures of the model classes involved in a system. The class diagram is one of the solutions that's used to design the document structure of the MongoDB database since there are no records, tables, or keys involved that are essential for ERD, like in a relational database. Designing a NoSQL database always requires an equal balance between the data retrieval methods and the data composition of the database. Data that will be stored in MongoDB always needs an ideal, feasible, and appropriate document structure, associations, aggregations, and layout. The following diagram shows the class diagram for our application's MongoDB database, `obrs`:

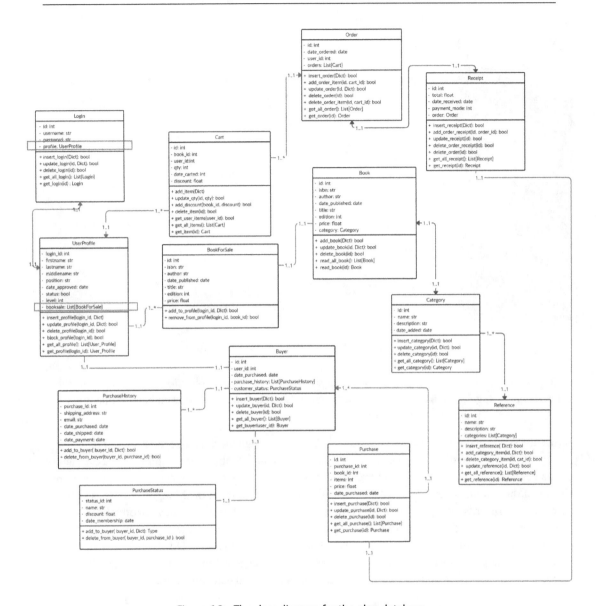

Figure 6.2 – The class diagram for the obrs database

Our application uses all the collections depicted in the preceding diagram to store all the information it captures from the client. Each context box represents one collection, with all the attributes and expected underlying transactions indicated inside the box. It also shows the associations that bind these collections, such as the one-to-one association between login and profile and the many-to-one association between BookForSale and UserProfile.

Now that the database server has been installed and designed, let us look at the different ways of establishing a connection from our FastAPI microservice application to its MongoDB database.

Applying the PyMongo driver for synchronous connections

We will start by learning how the FastAPI application connects to MongoDB using the PyMongo database driver. This driver is equivalent to `psycopg2`, which allows us to access PostgreSQL without using any ORM. Some popular ODMs, such as MongoEngine and Motor, use PyMongo as their core driver, which gives us the reason to explore PyMongo first before we touch on issues regarding popular ODMs. Studying the driver's behavior can provide baseline transactions that will show how an ODM builds the database connectivity, models, and CRUD transactions. But before we proceed with the details, we need to install the `pymongo` extension using `pip`:

```
pip install pymongo
```

Setting up the database connectivity

PyMongo uses its `MongoClient` module class to connect to any MongoDB database. We instantiate it with the specified host and port to extract the client object, such as `MongoClient("localhost", "27017")`, or a database URI, such as `MongoClient('mongodb://localhost:27017/')`. Our application uses the latter to connect to its database. But if we instantiate without providing the parameters, it will use the default `localhost` and `27017` details.

After extracting the client object, we can use it to access the database through a dot (`.`) operator or `attribute-style access` if the database name follows the Python naming convention; for example, `client.obrs`. Otherwise, we can use the bracket symbols (`[]`) or dictionary-style access; for example, `client["obrs_db"]`. Once the database object has been retrieved, we can access the collections using the access rules. Note that a collection is equivalent to a table in a relational database, where the collated records, known as documents, are stored. The following code shows a generator function that's used by the application to open database connectivity and access the necessary collections in preparation for the CRUD implementation:

```python
from pymongo import MongoClient

def create_db_collections():
    client = MongoClient('mongodb://localhost:27017/')
    try:
        db = client.obrs
        buyers = db.buyer
        users = db.login
        print("connect")
        yield {"users": users, "buyers": buyers}
```

```
finally:
    client.close()
```

A generator function such as `create_db_collections()` is preferred because the `yield` statement works perfectly when it comes to managing the database connection over the `return` statement. The `yield` statement suspends the function's execution when it sends a value back to the caller but retains the state where the function can resume at the point where it left off. This feature is applied by the generator to close the database connection when it resumes the execution at the `finally` clause. The `return` statement, on the other hand, will not be applicable for this purpose because `return` will finish the whole transaction before it sends a value to the caller.

However, before we invoke the generator, let us scrutinize how PyMongo builds its model layer to pursue the necessary CRUD transactions.

Building the model layer

Documents in MongoDB are represented and collated as JSON-style structures, specifically BSON documents. A BSON document offers more data types than the JSON structure. We can use dictionaries to represent and persist these BSON documents in PyMongo. Once a dictionary has been persisted, the BSON-type document will look like this:

```
{
    _id:ObjectId("61e7a49c687c6fd4abfc81fa"),
    id:1,
    user_id:10,
    date_purchased:"2022-01-19T00:00:00.000000",
    purchase_history:
    [
        {
            purchase_id:100,
            shipping_address:"Makati City",
            email:"mailer@yahoo.com",
            date_purchased:"2022-01-19T00:00:00.000000",
            date_shipped:"2022-01-19T00:00:00.000000",
            date_payment:"2022-01-19T00:00:00.000000"
        },
        {
            purchase_id:110,
            shipping_address:"Pasig City",
            email:"edna@yahoo.com",
```

```
            date_purchased:"2022-01-19T00:00:00.000000",
            date_shipped:"2022-01-19T00:00:00.000000",
            date_payment:"2022-01-19T00:00:00.000000"
        }
    ],
    customer_status:
    {
            status_id:90,
            name:"Sherwin John C. Tragura",
            discount:50,
            date_membership:"2022-01-19T00:00:00.000000"
    }
}
```

Common Python data types such as `str`, `int`, and `float` are supported by the BSON specification, but there are types such as `ObjectId`, `Decimal128`, `RegEx`, and `Binary` that are intrinsic only to the bson module. The specification only supports the `timestamp` and `datetime` temporal types. To install `bson`, use the following `pip` command:

```
pip install bson
```

> **Important note**
>
> **BSON** is short for **Binary JSON**, a serialized and binary encoding for JSON-like documents. The specification behind this is lightweight and flexible. The efficient encoding format is explained in more detail at `https://bsonspec.org/spec.html`.

`ObjectId` is an essential data type in a MongoDB document because it serves as a *unique identifier* for the main document structure. It is a *12-byte* field that consists of a 4-byte UNIX *embedded timestamp*, the 3-byte *machine ID* of the MongoDB server, a 2-byte *process ID*, and a 3-byte *arbitrary value* for the ID's increments. Conventionally, the declared field of the document, `_id`, always refers to the `ObjectId` value of the document structure. We can allow the MongoDB server to generate the `_id` object for the document or create an instance of the object type during persistence. When retrieved, `ObjectId` can be in *24 hexadecimal digit* or *string* format. Note that the `_id` field is the key indicator that a dictionary is ready to be persisted as a valid BSON document. Now, BSON documents can also be linked with one another using some associations.

Establishing document associations

MongoDB has no concept of referential integrity constraints, but a relationship among documents is possible based on structures. There are two types of documents: *main* and *embedded* documents. A document has a *one-to-one association* with another if it is an embedded document of the other. Likewise, a document has a *many-to-one association* if a list in that document is linked to the main document structure.

The previous purchase BSON document shows a sample of the principal `buyer` document with a one-to-one association with the `customer_status` embedded document and a many-to-one association with the `purchase_history` documents. As depicted from this sample document, embedded documents have no separate collection because they have no respective `_id` field to make them stand as primary documents.

Using the BaseModel classes for transactions

Since PyMongo has no predefined model classes, the Pydantic models of FastAPI can be used to represent MongoDB documents with all the necessary validation rules and encoders. We can use the `BaseModel` classes to contain document details and pursue *insert*, *update*, and *delete* transactions since the Pydantic models are compatible with MongoDB documents. The following models are being used by our online book reselling application to store and retrieve the `buyer`, `purchase_history`, and `customer_status` document details:

```python
from pydantic import BaseModel, validator
from typing import List, Optional, Dict
from bson import ObjectId
from datetime import date


class PurchaseHistoryReq(BaseModel):
    purchase_id: int
    shipping_address: str
    email: str
    date_purchased: date
    date_shipped: date
    date_payment: date

    @validator('date_purchased')
    def date_purchased_datetime(cls, value):
        return datetime.strptime(value,
            "%Y-%m-%dT%H:%M:%S").date()
```

```python
    @validator('date_shipped')
    def date_shipped_datetime(cls, value):
        return datetime.strptime(value,
            "%Y-%m-%dT%H:%M:%S").date()


    @validator('date_payment')
    def date_payment_datetime(cls, value):
        return datetime.strptime(value,
            "%Y-%m-%dT%H:%M:%S").date()


    class Config:
        arbitrary_types_allowed = True
        json_encoders = {
            ObjectId: str
        }


class PurchaseStatusReq(BaseModel):
    status_id: int
    name: str
    discount: float
    date_membership: date


    @validator('date_membership')
    def date_membership_datetime(cls, value):
        return datetime.strptime(value,
            "%Y-%m-%dT%H:%M:%S").date()


    class Config:
        arbitrary_types_allowed = True
        json_encoders = {
            ObjectId: str
        }


class BuyerReq(BaseModel):
```

```
_id: ObjectId
Buyer_id: int
user_id: int
date_purchased: date
purchase_history: List[Dict] = list()
customer_status: Optional[Dict]


@validator('date_purchased')
def date_purchased_datetime(cls, value):
    return datetime.strptime(value,
        "%Y-%m-%dT%H:%M:%S").date()


class Config:
    arbitrary_types_allowed = True
    json_encoders = {
        ObjectId: str
    }
```

For these request models to recognize the BSON data types, we should make some modifications to the default behavior of these models. Just like earlier in this chapter, where we added the orm_mode option, there is also a need to add a nested Config class to the BaseModel blueprint with the arbitrary_types_allowed option set to True. This additional configuration will recognize the BSON data types used in the attribute declaration, including compliance with the necessary underlying validation rules for the corresponding BSON data types used. Moreover, the json_encoders option should also be part of the configuration to convert the ObjectId property of the document into a string during a query transaction.

Using Pydantic validation

However, some other types are too complex for json_encoders to process, such as the BSON datettime field being converted into a Python datetime.date. Since the ODM cannot automatically convert a MongoDB datetime into a Python date type, we need to create a custom validation and parse this BSON datetime through Pydantic's @validation decorator. We must also use custom validators and parsers in the FastAPI services to convert all incoming Python date parameters into BSON datetime. This will be covered later.

`@validator` creates a `class` method that accepts `class` name as the first parameter, not the instance, of the field(s) to be validated and parsed. Its second parameter is an option that specifies the field name or class attribute that needs to be converted into another data type, such as `date_purchased`, `date_shipped`, or `date_payment` of the `PurchaseRequestReq` model. The `pre` attribute of `@validator` tells FastAPI to process the class methods before any built-in validation can be done in the API service implementation. These methods are executed right after `APIRouter` runs its custom and built-in FastAPI validation rules for the request models, if there are any.

Note that these request models have been placed in the `/models/request/buyer.py` module of the application.

Using the Pydantic @dataclass to query documents

Wrapping the queried BSON documents using the `BaseModel` model classes is still the best approach to implementing the query transaction. But since BSON has issues with the Python `datetime.date` fields, we cannot always utilize the request model classes that are used for the CRUD transaction by wrapping retrieved BSON documents. Sometimes, using the model yields an `"invalid date format (type=value_error.date)"` error because all the models have the Python `datetime.date` fields, whereas the incoming data has a BSON `datetime` or `timestamp`. Instead of adding more complexities to the request models, we should resort to another approach to extracting the documents – that is, utilizing the Pydantic `@dataclass`. The following data classes are defined for wrapping the extracted `buyer` documents:

```python
from pydantic.dataclasses import dataclass
from dataclasses import field
from pydantic import validator

from datetime import date, datetime
from bson import ObjectId
from typing import List, Optional

class Config:
        arbitrary_types_allowed = True

@dataclass(config=Config)
class PurchaseHistory:
    purchase_id: Optional[int] = None
```

```
        shipping_address: Optional[str] = None
        email: Optional[str] = None
        date_purchased: Optional[date] = "1900-01-01T00:00:00"
        date_shipped: Optional[date] = "1900-01-01T00:00:00"
        date_payment: Optional[date] = "1900-01-01T00:00:00"

        @validator('date_purchased', pre=True)
        def date_purchased_datetime(cls, value):
            return datetime.strptime(value,
                "%Y-%m-%dT%H:%M:%S").date()

        @validator('date_shipped', pre=True)
        def date_shipped_datetime(cls, value):
            return datetime.strptime(value,
                "%Y-%m-%dT%H:%M:%S").date()

        @validator('date_payment', pre=True)
        def date_payment_datetime(cls, value):
            return datetime.strptime(value,
                "%Y-%m-%dT%H:%M:%S").date()


@dataclass(config=Config)
class PurchaseStatus:
        status_id: Optional[int] = None
        name: Optional[str] = None
        discount: Optional[float] = None
        date_membership: Optional[date] = "1900-01-01T00:00:00"

        @validator('date_membership', pre=True)
        def date_membership_datetime(cls, value):
            return datetime.strptime(value,
                "%Y-%m-%dT%H:%M:%S").date()


@dataclass(config=Config)
class Buyer:
        buyer_id: int
```

```
    user_id: int
    date_purchased: date
    purchase_history: List[PurchaseHistory] =
        field(default_factory=list )
    customer_status: Optional[PurchaseStatus] =
        field(default_factory=dict)
    _id: ObjectId = field(default=ObjectId())


    @validator('date_purchased', pre=True)
    def date_purchased_datetime(cls, value):
        print(type(value))
        return datetime.strptime(value,
            "%Y-%m-%dT%H:%M:%S").date()
```

@dataclass is a decorator function that adds an __init__() to a Python class to initialize its attributes and other special functions, such as __repr__(). The PurchasedHistory, PurchaseStatus, and Buyer custom classes shown in the preceding code are typical classes that can be converted into request model classes. FastAPI supports both BaseModel and data classes when creating model classes. Apart from being under the Pydantic module, using @dataclass is not a replacement for using BaseModel when creating model classes. This is because the two components are different in terms of their flexibility, features, and hooks. BaseModel is configuration-friendly and can be adapted to many validation rules and type hints, while @dataclass has problems recognizing some Config attributes such as extra, allow_population_by_field_name, and json_encoders. If a data class requires some additional details, a custom class is needed to define these configurations and set the config parameter of the decorator. For instance, the Config class in the preceding code, which sets arbitrary_types_allowed to True, has been added to the three model classes.

Besides config, the decorator has other parameters such as init, eq, and repr that accept bool values to generate their respective hook methods. The frozen parameter enables exception handling concerning field type mismatches when set to True.

When it comes to data parsing, transition, and conversion, @dataclass is always dependent on augmented validations, unlike BaseModel, which can process data type conversion simply by adding json_encoders. In the data classes shown previously, all the validators focus on BSON datetime to Python datetime.date conversion during the document retrieval process. These validations will occur before any custom or built-in validation in APIRouter because the pre parameter of the @validator decorator is set to True.

When dealing with default values, BaseModel classes can use typical type hints such as Optional or object instantiation such as dict() or list() to define the preconditional state of its complex attributes. With @dataclass, a ValueError exception is always thrown at compile time when type hints are applied to set default values of complex field types such as list, dict, and ObjectId. It requires the field() specifier from Python's dataclasses module to set the default values of these fields, either by assigning an actual value through the specifier's default parameter or invoking a function or lambda that returns a valid value through the default_factory parameter. The use of field() indicates that Pydantic's @dataclass is an exact replacement of Python's core data classes but with some additional features, such as the config parameter and the inclusion of the @validator components.

Note that it is advised that all @dataclass models have default values when using type hints or field(), especially for embedded documents and for models with the date or datetime types, to avoid some missing constructor parameter(s) errors. On the other hand, an @dataclass can also create embedded structures in the BaseModel classes, for example, by defining attributes with the class types. This is highlighted in the Buyer model.

All these model classes have been placed in the /models/data/pymongo.py script. Let us now apply these data models to create the repository layer.

Implementing the repository layer

PyMongo needs collection to build the repository layer of the application. Besides the collection object, the *insert*, *delete*, and *update* transactions will also need the BaseModel classes to contain all the details from the client and convert them into BSON documents after the transaction. Meanwhile, our query transactions will require the data classes to convert all BSON documents into JSON-able resources during the document retrieval process. Now, let us look at how a repository can be implemented using a PyMongo driver.

Building the CRUD transactions

The repository class in the following code block implements the CRUD transactions that aim to manage the buyer, purchase_history, and customer_status information based on basic specifications of the *online book reselling* system:

```python
from typing import Dict, Any


class BuyerRepository:

    def __init__(self, buyers):
        self.buyers = buyers


    def insert_buyer(self, users,
```

```
        details:Dict[str, Any]) -> bool:
    try:
        user = users.find_one(
            {"_id": details["user_id"]})
        print(user)
        if user == None:
            return False
        else:
            self.buyers.insert_one(details)

    except Exception as e:
        return False
    return True
```

Let us examine insert_buyer(), which inserts details about a registered book buyer who had some previous transactions in the system as a login user. The PyMongo collection offers helper methods for processing CRUD transactions, such as insert_one(), which adds a single main document from its Dict parameter. It also has insert_many(), which accepts a valid list of dictionaries that can be persisted as multiple documents. These two methods can generate an ObjectId for the _id field of the BSON document during the insertion process. The buyer's details are extracted from the BuyerReq Pydantic model.

Next, update_buyer() shows how to update a specific document in the buyer collection:

```
def update_buyer(self, id:int,
        details:Dict[str, Any]) -> bool:
    try:
        self.buyers.update_one({"buyer_id": id},
            {"$set":details})
    except:
        return False
    return True

def delete_buyer(self, id:int) -> bool:
    try:
        self.buyers.delete_one({"buyer_id": id})
    except:
        return False
    return True
```

The collection has an `update_one()` method that requires two parameters: a unique and valid field/value dictionary pair that will serve as the *search key* of the record search, and another dictionary pair with the predefined `$set` key with the dictionary of updated *details for replacement*. It also has `update_many()`, which can update multiple documents, given that the primary dictionary field/value parameter is not unique.

`delete_buyer()` is the transaction that deletes a `buyer` document using a unique and valid field/value pair such as `{"buyer_id": id}`. If this parameter or search key is a common/non-unique data, the collection offers `delete_many()`, which can delete multiple documents. Now, the following script shows how to implement query transactions in PyMongo

```python
from dataclasses import asdict
from models.data.pymongo import Buyer
from datetime import datetime
from bson.json_util import dumps
import json

    ... ... ...

    ... ... ...

    ... ... ...

    def get_all_buyer(self):
        buyers = [asdict(Buyer(**json.loads(dumps(b))))
                for b in self.buyers.find()]
        return buyers


    def get_buyer(self, id:int):
        buyer = self.buyers.find_one({"buyer_id": id})
        return asdict(Buyer(**json.loads(dumps(buyer))))
```

When querying documents, PyMongo has a `find()` method, which retrieves all the documents in the collection, and `find_one()`, which can get a unique and single document. Both methods need two parameters: the conditional or logical query parameter in the form of a dictionary field/value pair and the set of fields that needs to appear in the record. `get_buyer()` in the previous code block shows how to retrieve a buyer document through the unique `buyer_id` field. The absence of its second parameter means the presence of all the fields in the result. Meanwhile, `get_all_buyer()` retrieves all the buyer documents without constraints. Constraints or filter expressions are formulated using BSON comparison operators, as shown in the following table:

Operator	Description
$eq	Equality between a field and a specified value
$gt	A field is greater than a specified value
$gte	A field is greater than or equal to a specified value
$in	A field is a part of an array of values
$lt	A field is less than a specified value
$lte	A field is less than or equal to a specified value
$ne	Non-equality between a field and a specified value
$nin	A field is not part of an array of values

For instance, retrieving buyer documents with *user_id greater than 5* requires the `buyers.find({"user_id": {"$gte": 5}})` query operation. If we need to build compound filters, we must apply the following logical operators:

Operator	Description
$not	Negates the value of the filter expression
$and	Evaluates to `True` if both filter expressions return `True`
$or	Evaluates to `True` if either of the filter expressions is `True`
$nor	Evaluates to `True` if both filter expressions return `False`

Retrieving buyer documents with *buyer_id less than 50* and *buyer_id greater than 10* will require the `find({'and': [{'buyer_id': {'$lt': 50}}, {'user_id':{'$gt':10}}]})` query.

Both methods return BSON documents that are not JSON-able components of the FastAPI framework. To convert the documents into JSON, the `bson.json_util` extension has a `dumps()` method that can convert a single document or list of documents into a JSON string. Both `get_all_buyer()` and `get_buyer()` convert every single document retrieved into JSON so that each can be mapped to the `Buyer` data class. The main objective of the mapping is to convert the `datetime` fields into Python `datetime.date` while utilizing the validators of the `Buyer` data class. The mapping will only be successful if the `loads()` method of the `json` extension is used to convert `str` into a `dict` data structure. After generating the list of `Buyer` data classes, the `asdict()` method of Python's `dataclasses` module is needed to transform the list of `Buyer` data classes into a list of dictionaries to be consumed by `APIRouter`.

Managing document association

Technically, there are two ways to construct a document association in PyMongo. The first one is to use the `DBRef` class of the `bison.dbref` module to link the parent and child documents. The only prerequisite is for both documents to have an `_id` value of the `ObjectId` type and have their respective collection exist. For instance, if `PurchaseHistoryReq` is a core document, we can insert one purchase record into the list through the following query:

```
buyer["purchase_history"].append(new  DBRef("purchase_history",
  "49a3e4e5f462204490f70911"))
```

Here, the first parameter of the DBRef constructor pertains to the name of the collection where the child document is placed, while the second one is the ObjectId property of the child document in string format. However, some people use an ObjectId instance instead of the string version. On the other hand, to find a specific purchase_history document from the buyer collection using DBRef, we can write our query like this:

```
buyer.find({ "purchase_history ": DBRef("purchase_
history",ObjectId("49a3e4e5f462204490f70911")) })
```

The second way is to add the whole BSON document structure to the list field of buyer through the BuyerReq model. This solution applies to embedded documents that do not have _id and collection but are essential to the core document. add_purchase_history() in the following code shows how this approach is applied to create a many-to-one association between the purchase_history and buyer documents:

```
def add_purchase_history(self, id:int,
                details:Dict[str, Any]):
    try:
        buyer = self.buyers.find_one({"buyer_id": id})
        buyer["purchase_history"].append(details)
        self.buyers.update_one({"buyer_id": id},
        {"$set": {"purchase_history":
                buyer["purchase_history"]}})
    except Exception as e:
        return False
    return True


def add_customer_status(self, id:int,
                details:Dict[str, Any]):
    try:
        buyer = self.buyers.find_one({"buyer_id": id})
        self.buyers.update_one({"buyer_id": id},
            {"$set":{"customer_status": details}})
    except Exception as e:
        return False
    return True
```

The add_customer_status() method shows how to implement the second approach in building a one-to-one association between the buyer and purchase_status documents. The first approach, which involves using DBRef, can also be applied if PurchaseStatusReq is an independent core document.

The complete repository class can be found in the /repository/pymongo/buyer.py script file. Now, let us apply these CRUD transactions to our API services.

Running the transactions

Before executing the BuyerRepository transactions, the create_db_collections() generator should be injected into the API services using Depends. Since PyMongo has difficulty processing Python types that are not BSON-supported, such as datettime.date, custom validations and serializers are sometimes required to pursue some transactions.

> **Important note**
> The implementation of @validator inside @dataclass and BaseModel converts outgoing BSON datetime parameters into Python date during query retrieval. Meanwhile, the JSON encoder validation in this API layer converts incoming Python date values into BSON datetime values during the transition from the application to MongoDB.

For instance, the add_buyer(), update_buyer(), and add_purchase_history() transaction methods in the following code require a custom serializer such as json_serialize_date() to transform the Python datetime.date value into the datettime.datetime type so that it complies with PyMongo's BSON specification:

```
from fastapi import APIRouter, Depends
from fastapi.responses import JSONResponse

from models.request.buyer import BuyerReq,
        PurchaseHistoryReq, PurchaseStatusReq
from repository.pymongo.buyer import BuyerRepository
from db_config.pymongo_config import create_db_collections

from datetime import date, datetime
from json import dumps, loads
from bson import ObjectId


router = APIRouter()
```

```python
def json_serialize_date(obj):
    if isinstance(obj, (date, datetime)):
        return obj.strftime('%Y-%m-%dT%H:%M:%S')
    raise TypeError ("The type %s not serializable." %
            type(obj))


def json_serialize_oid(obj):
    if isinstance(obj, ObjectId):
        return str(obj)
    elif isinstance(obj, date):
        return obj.isoformat()
    raise TypeError ("The type %s not serializable." %
            type(obj))


@router.post("/buyer/add")
def add_buyer(req: BuyerReq,
            db=Depends(create_db_collections)):
    buyer_dict = req.dict(exclude_unset=True)
    buyer_json = dumps(buyer_dict,
              default=json_serialize_date)
    repo:BuyerRepository = BuyerRepository(db["buyers"])
    result = repo.insert_buyer(db["users"],
            loads(buyer_json))

    if result == True:
        return JSONResponse(content={"message":
          "add buyer successful"}, status_code=201)
    else:
        return JSONResponse(content={"message":
          "add buyer unsuccessful"}, status_code=500)


@router.patch("/buyer/update")
def update_buyer(id:int, req:BuyerReq,
            db=Depends(create_db_collections)):
    buyer_dict = req.dict(exclude_unset=True)
    buyer_json = dumps(buyer_dict,
```

```
                  default=json_serialize_date)
    repo:BuyerRepository = BuyerRepository(db["buyers"])
    result = repo.update_buyer(id, loads(buyer_json))

    if result == True:
        return JSONResponse(content={"message":
          "update buyer successful"}, status_code=201)
    else:
        return JSONResponse(content={"message":
          "update buyer unsuccessful"}, status_code=500)


@router.post("/buyer/history/add")
def add_purchase_history(id:int, req:PurchaseHistoryReq,
          db=Depends(create_db_collections)):
    history_dict = req.dict(exclude_unset=True)
    history_json = dumps(history_dict,
          default=json_serialize_date)
    repo:BuyerRepository = BuyerRepository(db["buyers"])
    result = repo.add_purchase_history(id,
          loads(history_json))
```

The `json_serialize_date()` function becomes part of the JSON serialization process of the `dumps()` method but only handles the temporal type conversion while transforming the `buyer` details into JSON objects. It is applied in the *INSERT* and *UPDATE* transactions of the repository class to extract the serialized JSON string equivalent of the `BuyerReq`, `PurchaseHistoryReq`, and `PurchaseStatusReq` models.

Now, another custom converter is applied in the data retrievals of the `list_all_buyer()` and `get_buyer()` methods:

```
@router.get("/buyer/list/all")
def list_all_buyer(db=Depends(create_db_collections)):
    repo:BuyerRepository = BuyerRepository(db["buyers"])
    buyers = repo.get_all_buyer()
    return loads(dumps(buyers, default=json_serialize_oid))


@router.get("/buyer/get/{id}")
def get_buyer(id:int, db=Depends(create_db_collections)):
```

```
repo:BuyerRepository = BuyerRepository(db["buyers"])
buyer = repo.get_buyer(id)
return loads(dumps(buyer, default=json_serialize_oid))
```

The data models involved in our query transactions are data classes, so the results of the two preceding query methods have already been mapped and transformed into JSON format. However, unfortunately, they're not JSON-able enough for the FastAPI framework. Aside from BSON datetime types, the PyMongo ODM cannot automatically convert ObjectId into a default type in Python, thus throwing ValueError during data retrieval from MongoDB. To fix this problem, dumps() needs a custom serializer, such as json_serialize_oid(), to convert all ObjectId parameters in MongoDB into FastAPI transitions. It also converts BSON datetime values into Python date values following the *ISO-8601* format. The valid JSON string from dumps() will enable the loads() method to produce a JSON-able result for the FastAPI services. The complete API services can be found in the /api/buyer.py script file.

After complying with all the requirements, PyMongo can help store and manage all the information using the MongoDB server. However, the driver only works for synchronous CRUD transactions. If we opt for an asynchronous way of implementing CRUD, we must always resort to the Motor driver.

Creating async CRUD transactions using Motor

Motor is an asynchronous driver that relies on the AsyncIO environment of the FastAPI. It wraps PyMongo to produce non-blocking and coroutine-based classes and methods needed to create asynchronous repository layers. It is almost like PyMongo when it comes to most of the requirements except for the database connectivity and repository implementation.

But before we proceed, we need to install the motor extension using the following pip command:

```
pip install motor
```

Setting up the database connectivity

Using the AsyncIO platform of the FastAPI, the Motor driver opens a connection to the MongoDB database through its AsyncIOMotorClient class. When instantiated, the default connection credential is always localhost at port 27017. Alternatively, we can specify the new details in str format through its constructor. The following script shows how to create a global AsyncIOMotorClient reference with the specified database credentials:

```
from motor.motor_asyncio import AsyncIOMotorClient


def create_async_db():
    global client
    client = AsyncIOMotorClient(str("localhost:27017"))
```

```
def create_db_collections():
    db = client.obrs
    buyers = db["buyer"]
    users = db["login"]
    return {"users": users, "buyers": buyers}

def close_async_db():
    client.close()
```

The format of the database URI is a string with a colon (:) in between the details. Now, the application needs the following Motor methods to start the database transactions:

- `create_async_db()`: A method for establishing the database connection and loading schema definitions

- `close_async_db()`: A method for closing the connection

`APIRouter` will require event handlers to manage these two core methods as application-level events. Later, we will register `create_async_db()` as a startup event and `close_async_db()` as a shutdown event. On the other hand, the `create_db_collections()` method creates some references to the `login` and `buyer` collections, which will be needed by the repository transactions later.

In general, creating the database connection and getting the reference to the document collections do not require the `async/await` expression since no I/O is involved in the process. These methods can be found in the `/db_config/motor_config.py` script file. It is time now to create Motor's repository layer.

Creating the model layer

PyMongo and Motor share the same approaches in creating both the request and data models. All base models, data classes, validators, and serializers used by PyMongo also apply to Motor connectivity.

Building the asynchronous repository layer

When it comes to the CRUD implementation, both PyMongo and Motor have some slight differences in the syntax but a considerable difference in the performance of each transaction. Their helper methods for inserting, updating, and deleting documents, including the necessary method parameters, are all the same, except that Motor has the non-blocking versions. Invoking the non-blocking Motor methods inside the repository requires an async/await expression. Here is an asynchronous version of PyMongo's `BuyerRepository`:

```
class BuyerRepository:
```

```
def __init__(self, buyers):
    self.buyers = buyers


async def insert_buyer(self, users,
        details:Dict[str, Any]) -> bool:
    try:
        user = await users.find_one({"_id":
            details["user_id"]})
        … … … … …
        else:
            await self.buyers.insert_one(details)
        … … … … …

    return True

async def add_purchase_history(self, id:int,
        details:Dict[str, Any]):
    try:
        … … … … …
        await self.buyers.update_one({"buyer_id": id},
            {"$set":{"purchase_history":
                buyer["purchase_history"]}})
        … … … … …
    return True
```

insert_buyer() in the preceding code block is defined as async because insert_one() is a non-blocking operation that requires an await invocation. The same goes for add_purchase_history(), which updates the purchase_history embedded documents using the non-blocking update_one():

```
async def get_all_buyer(self):
    cursor = self.buyers.find()
    buyers = [asdict(Buyer(**json.loads(dumps(b))))
        for b in await cursor.to_list(length=None)]
    return buyers

async def get_buyer(self, id:int):
```

```
          buyer = await self.buyers.find_one(
                      {"buyer_id": id})
          return asdict(Buyer(**json.loads(dumps(buyer))))
```

The delete_many() and find_one() operations are also invoked through an await expression. However, find() in Motor is not asynchronous and behaves differently than it does with PyMongo. The reason is that find() is not an I/O operation in Motor, and it returns an AsyncIOMotorCursor or asynchronous cursor, an iterable type that contains all the BSON documents. We apply async to the cursor when retrieving all its stored documents. The get_all_buyer() transaction in the preceding code shows how we call the find() operation and invoke the cursor to extract the necessary documents for JSON transformation. This repository class can be found in the /repository/motor/buyer.py script file. Let us now apply these CRUD transactions to our API services.

Running the CRUD transactions

For the repository to work with APIRouter, we need to create two event handlers to manage the database connection and document collection retrieval. The first event, which is the startup event that the Uvicorn server executes before the application runs, should trigger the create_async_db() method's execution to instantiate AsyncIOMotorClient and make references to the collections. The second event, which is the shutdown event, runs when the Uvicorn server is shutting down and should trigger the close_async_db() execution to close the connection. APIRouter has an add_event_handler() method to create these two event handlers. The following is a portion of the APIRouter script that shows how to prepare the database connection for the BuyerRepository transactions:

```
... ... ... ... ... ...
from db_config.motor_config import create_async_db,
    create_db_collections, close_async_db

... ... ... ... ... ...

router = APIRouter()

router.add_event_handler("startup",
            create_async_db)
router.add_event_handler("shutdown",
            close_async_db)
```

The "startup" and "shutdown" values are pre-built configuration values and not just any arbitrary string values used to indicate the type of event handlers. We will discuss these event handlers in more detail in *Chapter 8, Creating Coroutines, Events, and Message-Driven Transactions*.

After setting these events handlers, the API services can now invoke the repository transactions asynchronously using an await/async expression. The validations and serialization utilities that are applied in PyMongo can also be utilized here in this version of `BuyerRepository`. The collections will be available to the API services upon injecting `create_db_collections()` into the API services. The `add_buyer()` API service showcases the implementation of an asynchronous REST transaction using the Motor driver:

```
@router.post("/buyer/async/add")
async def add_buyer(req: BuyerReq,
        db=Depends(create_db_collections)):
    buyer_dict = req.dict(exclude_unset=True)
    buyer_json = dumps(buyer_dict,
            default=json_serialize_date)
    repo:BuyerRepository = BuyerRepository(db["buyers"])

    result = await repo.insert_buyer(db["users"],
            loads(buyer_json))
    if result == True:
        return JSONResponse(content={"message":
            "add buyer successful"}, status_code=201)
    else:
        return JSONResponse(content={"message":
            "add buyer unsuccessful"}, status_code=500)
```

Using PyMongo and Mongo drivers provides a minimal and exhaustive implementation of the MongoDB transactions. The core implementation of every CRUD transaction varies from one developer to another, and the approaches that are used to scrutinize and analyze the processes involved are managed in different ways. Also, there are no established standards for defining the document fields, such as *data uniqueness, the length of the field value, the value range*, and even the idea of adding a *unique ID*. To address these issues surrounding PyMongo and Motor, let us explore other ways of opening a connection to MongoDB to create CRUD transactions, such as using an **ODM**.

Implementing CRUD transactions using MongoEngine

MongoEngine is an ODM that uses PyMongo to create an easy-to-use framework that can assist in managing MongoDB documents. It offers API classes that can help generate model classes using its field types and attribute metadata. It provides a declarative way of creating and structuring the embedded documents.

Before we explore this ODM, we need to install it using the following `pip` command:

```
pip install mongoengine
```

Establishing database connection

MongoEngine has one of the most straightforward ways to establish a connection. Its mongoengine module has a `connect()` helper method that connects to the MongoDB database when it's given the appropriate database connections. Our application must have a generator method to create a reference to the database connection and close this created connection after the transactions expire. The following script showcases the MongoEngine database connectivity:

```python
from mongoengine import connect

def create_db():
    try:
        db = connect(db="obrs", host="localhost",
                port=27017)
        yield db
    finally:
        db.close()
```

The `connect()` method has a mandatory first parameter, named db, which indicates the name of the database. The remaining parameters refer to the other remaining details of the database connection, such as `host`, `port`, `username`, and `password`. This configuration can be found in the `/db_config/mongoengine_config.py` script file. Let us now create data models for our MongoEngine repository.

Building the model layer

MongoEngine provides a convenient and declarative way of mapping BSON documents to the model classes through its Document API class. A model class must subclass Document to inherit the structure and properties of a qualified and valid MongoDB document. The following is a Login definition that's been created using the Document API class:

```python
from mongoengine import Document, StringField,
        SequenceField, EmbeddedDocumentField
import json

class Login(Document):
    id = SequenceField(required=True, primary_key=True)
```

```
username = StringField(db_field="username",
    max_length=50, required=True, unique=True)
password = StringField(db_field="password",
    max_length=50, required=True)
profile = EmbeddedDocumentField(UserProfile,
    required=False)


def to_json(self):
        return {
        "id": self.id,
        "username": self.username,
        "password": self.password,
        "profile": self.profile
    }


@classmethod
def from_json(cls, json_str):
    json_dict = json.loads(json_str)
    return cls(**json_dict)
```

Unlike PyMongo and the Motor drivers, MongoEngine can define class attributes using its `Field` classes and their properties. Some of its `Field` classes include `StringField`, `IntField`, `FloatField`, `BooleanField`, and `DateField`. These can declare the `str`, `int`, `float`, `bool`, and `datetime.date` class attributes, respectively.

Another convenient feature that this ODM has is that it can create `SequenceField`, which behaves the same as the `auto_increment` column field in a relational database or `Sequence` in an object-relational database. The `id` field of a model class should be declared as `SequenceField` so that it serves as the primary key of the document. Like in a typical sequence, this field has utilities to increment its value or reset it to zero, depending on what document record must be accessed.

Other than the field types, field classes can also provide field arguments to attributes such as `choices`, `required`, `unique`, `min_value`, `max_value`, `max_length`, and `min_length` to give constraints to the field values. The `choices` parameter, for instance, accepts an iterable of string values that will serve as an enumeration. The `required` parameter indicates whether the field always needs a field value, while the `unique` parameter means the field value has no duplicates in the collection. Violating the `unique` parameter will lead to the following error message:

```
Tried to save duplicate unique keys (E11000 duplicate key error
collection: obrs.login index: username_ ...)
```

`min_value` and `max_value`, on the other hand, indicate the minimum and maximum values for the numeric fields, respectively. `min_length` specifies the minimum length of a string value, while `max_length` sets the maximum string length. The `db_field` parameter, on the other hand, can also be applied when specifying another document field name instead of the class attribute name. The given `Login` class also has `username` and `password` fields defined to hold string values, an id primary key defined as `SequenceField`, and an embedded document field to establish document association.

Creating document association

The `profile` field of `Login` creates a one-to-one association between the `Login` document and `UserProfile`. But before the association can work, we need to define the `profile` field as being of the `EmbeddedDocumentField` type and `UserProfile` as being of the `EmbeddedDocument` type. The following is the complete blueprint of `UserProfile`:

```
class UserProfile(EmbeddedDocument):
    firstname = StringField(db_field="firstname",
            max_length=50, required=True)
    lastname = StringField(db_field="lastname",
            max_length=50, required=True)
    middlename = StringField(db_field="middlename",
            max_length=50, required=True)
    position = StringField(db_field="position",
            max_length=50, required=True)
    date_approved = DateField(db_field="date_approved",
            required=True)
    status = BooleanField(db_field="status", required=True)
    level = IntField(db_field="level", required=True)
    login_id = IntField(db_field="login_id", required=True)
    booksale = EmbeddedDocumentListField(BookForSale,
            required=False)

    def to_json(self):
            return {
            "firstname": self.firstname,
            "lastname": self.lastname,
            "middlename": self.middlename,
            "position": self.position,
            "date_approved":
```

```
                self.date_approved.strftime("%m/%d/%Y"),
            "status": self.status,
            "level": self.level,
            "login_id": self.login_id,
            "books": self.books
        }

    @classmethod
    def from_json(cls, json_str):
        json_dict = json.loads(json_str)
        return cls(**json_dict)
```

The EmbeddedDocument API is a Document without an id and has no collection of its own. Subclasses of this API are model classes that have been created to be part of a core document structure, such as UserProfile being part of the Login details. Now, the field that refers to this document has a required property set to False since an embedded document can't be present at all times.

On the other hand, a field declared as EmbeddedDocumentList is used to create a many-to-one association between documents. The preceding UserProfile class is strongly connected to a list of BookForSale embedded documents because of its declared booksale field. Again, the field type should always set its required property to False to avoid problems when dealing with empty values.

Applying custom serialization and deserialization

There are no built-in hooks for validation and serialization in this ODM. Every model class in the *online book reselling* application has implemented a from_json() class method that converts JSON details into a valid Document instance. When converting the BSON document into a JSON object, model classes must have the custom to_json() instance method, which builds the JSON structure and automatically transforms the BSON datetime into JSON-able date objects through formatting. Let us now create the repository layer using the model classes.

Implementing the CRUD transactions

MongoEngine provides the most convenient and straightforward approach to building the repository layer for the application. All its operations come from the Document model class and they are easy to use. LoginRepository uses the ODM to implement its CRUD transactions:

```
from typing import Dict, Any
from models.data.mongoengine import Login


class LoginRepository:
```

```python
def insert_login(self, details:Dict[str, Any]) -> bool:
    try:
        login = Login(**details)
        login.save()
    except Exception as e:
        print(e)
        return False
    return True


def update_password(self, id:int, newpass:str) -> bool:
    try:
        login = Login.objects(id=id).get()
        login.update(password=newpass)
    except:
        return False
    return True


def delete_login(self, id:int) -> bool:
    try:
        login = Login.objects(id=id).get()
        login.delete()
    except:
        return False
    return True
```

It only takes two lines for the insert_login() method to save the Login document. After creating the Login instance with the necessary document details, we simply call the save() method of the Document instance to pursue the insert transaction. When it comes to modifying some document values, the Document API class has an update() method that manages changes in state for every class attribute. But first, we need to find the document using the objects() utility method, which retrieves document structures from the collection. This objects() method can fetch a document by providing its parameter with an id field value or extracting a list of document records by supplying the method with a generic search expression. The instance of the retrieved document must invoke its update() method to pursue the modification of some, if not all, of its field values. The given update_password() method updates the password field of Login, which gives us a good template regarding how to pursue update operations on other field attributes.

On the other hand, `delete_login()` shows how to delete a `Login` document from its collection after it searches for the object using a simple call to the instance's `delete()` method. The following script shows how to perform query transactions in MongoEngine:

```python
def get_all_login(self):
    login = Login.objects()
    login_list = [l.to_json() for l in login]
    return login_list

def get_login(self, id:int):
    login = Login.objects(id=id).get()
    return login.to_json()
```

The only way to perform single- or multiple-document retrieval is to utilize the `objects()` method. There is no need to implement JSON converters for the query results because every `Document` model class has a `to_json()` method to provide the JSON-able equivalent of the instance. The given `get_all_login()` transaction uses list comprehension to create a list of JSON documents from the result of `objects()`, while the `get_login()` method invokes `to_json()` after extracting a single document.

Managing the embedded documents

It is easier to implement document associations with an ODM than the core PyMongo and Motor database drivers. Since the operations of MongoEngine are comfortable to use, it takes only a few lines to manage the embedded documents. In the following `UserProfileRepository` script, `insert_profile()` shows how adding a `UserProfile` detail to the `Login` document can be done by performing a simple object search and an `update()` call:

```python
from typing import Dict, Any
from models.data.mongoengine import Login, UserProfile,
    BookForSale

class UserProfileRepository():

    def insert_profile(self, login_id:int,
            details:Dict[str, Any]) -> bool:
        try:
            profile = UserProfile(**details)
            login = Login.objects(id=login_id).get()
            login.update(profile=profile)
```

```
        except Exception as e:
            print(e)
            return False
        return True


    def add_book_sale(self, login_id:int,
            details:Dict[str, Any]):
        try:
            sale = BookForSale(**details)
            login = Login.objects(id=login_id).get()
            login.profile.booksale.append(sale)
            login.update(profile=login.profile)
        except Exception as e:
            print(e)
            return False
        return True
```

Likewise, the given `add_book_sale()` transaction creates a many-to-one association between `BookForSale` and `UserProfile` using the same approach applied in `insert_profile()` with an additional List's `append()` operation.

Querying the embedded documents is also feasible in MongoEngine. The ODM has a `filter()` method that uses *field lookup syntax* to refer to a specific document structure or list of embedded documents. This field lookup syntax consists of the *field name of the embedded document*, followed by a *double underscore* in place of the dot in the usual object attribute access syntax. Then, it has *another double underscore* to cater to some *operators*, such as `lt`, `gt`, `eq`, and `exists`. In the following code, `get_all_profile()` uses the `profile__login_id__exists=True` field lookup to filter all `user_profile` embedded documents that have valid `login` structures. However, the `get_profile()` transaction does not need to use `filter()` and field lookups because it can simply access the specific login document to fetch its profile details:

```
    def get_all_profile(self):
        profiles = Login.objects.filter(
                profile__login_id__exists=True)
        profiles_dict = list(
            map(lambda h: h.profile.to_json(),
                Login.objects().filter(
                    profile__login_id__exists=True)))
        return profiles_dict
```

```
def get_profile(self, login_id:int):
    login = Login.objects(id=login_id).get()
    profile = login.profile.to_json()
    return profile
```

The preceding query transactions are just simple implementations compared to some other complex MongoEngine queries, which involve complicated embedded document structures that require complex field lookup syntax. Let us now apply the CRUD transactions to our API services.

Running the CRUD transactions

CRUD will not work without passing our `create_db()` method to the *startup* event and `disconnect_db()` to the *shutdown* event. The former will open the MongoDB connection during the Uvicorn startup, while the latter will close it during server shutdown.

The following script shows the application's `profile` router with a `create_profile()` REST service that asks clients for a profile detail, given a specific login record, and pursues the insert transaction using `UserProfileRepository`:

```python
from fastapi import APIRouter, Depends
from fastapi.responses import JSONResponse


from models.request.profile import UserProfileReq,
        BookForSaleReq
from repository.mongoengine.profile import
        UserProfileRepository
from db_config.mongoengine_config import create_db


router = APIRouter()


@router.post("/profile/login/add",
        dependencies=[Depends(create_db)])
def create_profile(login_id:int, req:UserProfileReq):
    profile_dict = req.dict(exclude_unset=True)
    repo:UserProfileRepository = UserProfileRepository()
    result = repo.insert_profile(login_id, profile_dict)
    if result == True:
        return req
    else:
```

```
        return JSONResponse(content={"message":
          "insert profile unsuccessful"}, status_code=500)
```

`create_profile()` is a standard API service that deals with MongoEngine's synchronous `insert_profile()` transaction. When it comes to asynchronous REST services, it is not advisable to use MongoEngine because its platform only works with synchronous ones. In the next section, we will discuss an ODM that's popular in building an asynchronous repository layer.

Implementing async transactions using Beanie

Beanie is a non-boilerplate mapper that utilizes the core features of Motor and Pydantic. This ODM offers a more straightforward approach to implementing asynchronous CRUD transactions than its precursor, the Motor driver.

To use Beanie, we need to install it using the following `pip` command:

```
pip install beanie
```

> **Important note**
> Installing Beanie may uninstall the current version of your Motor module because it sometimes requires lower version of Motor module. Pursuing this will produce errors in your existing Motor transactions.

Creating the database connection

Beanie uses the Motor driver to open a database connection to MongoDB. Instantiating the Motor's `AsyncIOMotorClient` class with the database URL is the first step of configuring it. But what makes Beanie unique compared to other ODMs is how it pre-initializes and pre-recognizes the model classes that will be involved in a CRUD transaction. The ODM has an asynchronous `init_beanie()` helper method that initiates the model class initialization using the database name. Calling this method will also set up the collection-domain mapping, where all the model classes are registered in the `document_models` parameter of `init_beanie()`. The following script shows the database configuration that's required to access our MongoDB database, `obrs`:

```
from motor.motor_asyncio import AsyncIOMotorClient
from beanie import init_beanie
from models.data.beanie import Cart, Order, Receipt


async def db_connect():
    global client
    client =
```

```
    AsyncIOMotorClient(f"mongodb://localhost:27017/obrs")
    await init_beanie(client.obrs,
        document_models=[Cart, Order, Receipt])


async def db_disconnect():
    client.close()
```

Here, db_connect() uses an async/await expression because its method invocation to init_beanie() is asynchronous. db_disconnect() will close the database connection by calling the close() method of the AsyncIOMotorClient instance. Both of these methods are executed as events, just like in MongoEngine. Their implementation can be found in the /db_config/beanie_config.py script file. Let us now create the model classes.

Defining the model classes

The Beanie ODM has a Document API class that's responsible for defining its model classes, mapping them to MongoDB collections, and handling repository transactions, just like in MongoEngine. Although there is no Field directive for defining class attributes, the ODM supports Pydantic's validation and parsing rules and typing extension for declaring models and their attributes. But it also has built-in validation and encoding features, which can be used together with Pydantic. The following script shows how to define Beanie model classes while it's being configured:

```
from typing import Optional, List
from beanie import Document
from bson import datetime


class Cart(Document):
    id: int
    book_id: int
    user_id: int
    qty: int
    date_carted: datetime.datetime
    discount: float

    class Collection:
        name = "cart"

    ... ... ... ... ... ...


class Order(Document):
```

```
    id: int
    user_id: int
    date_ordered: datetime.datetime
    orders: List[Cart] = list()


    class Collection:
        name = "order"
    ... ... ... ... ... ...


class Receipt(Document):
    id: int
    date_receipt: datetime.datetime
    total: float
    payment_mode: int
    order: Optional[Order] = None


    class Collection:
        name = "receipt"
    class Settings:
        use_cache = True
        cache_expiration_time =
            datetime.timedelta(seconds=10)
        cache_capacity = 10
```

The id attribute of the given Document classes automatically translates into an _id value. This serves as the primary key of the document. Beanie allows you to replace the default ObjectId type of _id with another type, such as int, which is not possible in other ODMs. And with Motor, this ODM needs custom JSON serializers because it has difficulty converting BSON datetime types into Python datetime.date types during CRUD transactions.

A document in Beanie can be configured by adding the Collection and Settings nested classes. The Collection class can replace the default name of the collection where the model is supposed to be mapped. It can also provide indexes to document fields if needed. The Settings inner class, on the other hand, can override existing BSON encoders, apply caching, manage concurrent updates, and add validation when the document is being saved. These three model classes include the collection configuration in their definitions to replace the names of their respective collections with their class names.

Creating the document associations

Python syntax, Pydantic rules, and API classes are used to establish links between documents in this mapper. To create a one-to-one association between `Order` and `Receipt`, for instance, we only need to set an `Order` field attribute that will link to a single `Receipt` instance. For many-to-one associations, such as the relationship between `Order` and `Cart`, the `Cart` document should only need a list field that will contain all the `Order` embedded documents.

However, the ODM has a `Link` type, which can be used to define class fields to generate these associations. Its CRUD operations, such as `save()`, `insert()`, and `update()`, strongly support these `Link` types, so long as the `link_rule` parameter is provided in their parameters. For query transactions, the `find()` method can include the `Link` documents during document fetching, given that its `fetch_links` parameter is set to `True`. Now, let us implement the repository layer using the model classes.

Implementing the CRUD transactions

Implementing repositories with Beanie is similar to how it's done with MongoEngine – that is, it uses short and direct CRUD syntax due to the convenient helper methods like create(), update(), and delete(), provided by the `Document` API class. However, the Beanie mapper creates an asynchronous repository layer because all the API methods that are inherited by the model classes are non-blocking. The following code for the `CartRepository` class shows a sample implementation of an asynchronous repository class using this Beanie ODM:

```python
from typing import Dict, Any
from models.data.beanie import Cart


class CartRepository:

    async def add_item(self,
            details:Dict[str, Any]) -> bool:
        try:
            receipt = Cart(**details)
            await receipt.insert()
        except Exception as e:
            print(e)
            return False
        return True

    async def update_qty(self, id:int, qty:int) -> bool:
        try:
```

```
            cart = await Cart.get(id)
            await cart.set({Cart.qty:qty})
        except:
            return False
        return True


    async def delete_item(self, id:int) -> bool:
        try:
            cart = await Cart.get(id)
            await cart.delete()
        except:
            return False
        return True
```

The add_item() method showcases the use of the asynchronous insert() method to persist a newly created Cart instance. The Document API also has a create() method that works like insert(). Another option is to use the insert_one() class method instead of the instance methods. Moreover, adding multiple documents is allowed in this ODM because an insert_many() operation exists to pursue that kind of insert.

Updating a document can be initiated using two methods, namely set() and replace(). update_qty() in the preceding script chooses the set() operation to update the current qty value of the items placed in a cart.

When it comes to document removal, the ODM only has the delete() method to pursue the transactions. This is present in the delete_item() transaction in the preceding code.

Retrieving a single document or a list of documents using this ODM is easy. No further serialization and cursor wrapping is needed during its query operations. When fetching a single document structure, the mapper provides the get() method if the fetching process only requires the _id field; it provides find_one() when the fetching process requires a conditional expression. Moreover, Beanie has a find_all() method that fetches all the documents without constraints and the find() method for retrieving data with conditions. The following code shows the query transaction for retrieving cart items from the database:

```
async def get_cart_items(self):
        return await Cart.find_all().to_list()


    async def get_items_user(self, user_id:int):
        return await Cart.find(
            Cart.user_id == user_id).to_list()
```

```
async def get_item(self, id:int):
    return await Cart.get(id)
```

Both the `find()` and `find_all()` operations are used in the methods to return a `FindMany` object that has a `to_list()` utility that returns a list of JSON-able documents. Let us now apply our CRUD transactions to the API services.

Running the repository transactions

The `CartRepository` methods will only run successfully if `db_connect()` from the configuration file is injected into the router. Although injecting it into each API service is acceptable, our solution prefers injecting the component into `APIRouter` using `Depends`:

```
from repository.beanie.cart import CartRepository
from db_config.beanie_config import db_connect
router = APIRouter(dependencies=[Depends(db_connect)])
@router.post("/cart/add/item")
async def add_cart_item(req:CartReq):
    repo:CartRepository = CartRepository()
    result = await repo.add_item(loads(cart_json))
            "insert cart unsuccessful"}, status_code=500)
```

The asynchronous `add_cart_item()` service asynchronously inserts the cart account into the database using `CartRepository`.

Another asynchronous mapper that can integrate perfectly with FastAPI is *ODMantic*.

Building async repository for FastAPI using ODMantic

The dependencies of Beanie and ODMantic come from Motor and Pydantic. ODMantic also utilizes Motor's `AsyncIOMotorClient` class to open a database connection. It also uses Pydantic features for class attribute validation, Python's typing extension for type hinting, and other Python components for management. But its edge over Beanie is that it complies with ASGI frameworks such as FastAPI.

To pursue ODMantic, we need to install the extension using the following `pip` command:

```
pip install odmantic
```

Creating the database connection

Setting up the database connectivity in ODMantic is the same as what we do with the Beanie mapper, except that the setup includes creating an engine that will handle all its CRUD operations. This engine is `AIOEngine` from the `odmantic` module, which requires both the motor client object and the database name to be created successfully. The following is a complete implementation of the database connectivity needed by the ODMantic mapper:

```python
from odmantic import AIOEngine
from motor.motor_asyncio import AsyncIOMotorClient

def create_db_connection():
    global client_od
    client_od =
        AsyncIOMotorClient(f"mongodb://localhost:27017/")

def create_db_engine():
    engine = AIOEngine(motor_client=client_od,
            database="obrs")
    return engine

def close_db_connection():
    client_od.close()
```

We need to create event handlers in `APIRouter` to run `create_db_connection()` and `close_db_connection()` for our repository transactions to work. Let us now implement the model layer of the ODM.

Creating the model layer

ODMantic has a `Model` API class that provides properties to model classes when subclassed. It relies on Python types and BSON specifications to define the class attributes. When transforming field types, such as converting a BSON `datetime` value into a Python `datetime.date` value, the mapper allows you to add custom `@validator` methods into the model classes to implement the appropriate object serializer. Generally, ODMantic relies on the `pydantic` module when it comes to data validation, unlike in the Beanie mapper. The following is a standard ODMantic model class definition:

```python
from odmantic import Model
from bson import datetime
```

```
class Purchase(Model):
    purchase_id: int
    buyer_id: int
    book_id: int
    items: int
    price: float
    date_purchased: datetime.datetime

    class Config:
        collection = "purchase"
```

For advanced configurations, we can add a nested `Config` class to the model class to set these additional options, such as the `collection` option, which replaces the default name of the collection with a custom one. We can also configure some familiar options, such as `json_encoders`, to convert one field type into another supported one.

Establishing document association

When creating associations, the typical Python approach of declaring fields so that they refer to an embedded document(s) is still applicable in this ODM. However, this ODM mapper has an `EmbeddedModel` API class to create a model with no `_id` field; this can be linked to another document. The `Model` classes, on the other hand, can define a field attribute that will refer to an `EmbeddedModel` class to establish a one-to-one association or a list of `EmbeddedModel` instances for a many-to-one association.

Implementing the CRUD transactions

Creating the repository layer using ODMantic always requires the engine object that was created in the startup event. This is because all the CRUD operations that are needed will come from this engine. The following `PurchaseRepository` shows the operations from the `AIOEngine` object that we need to create CRUD transactions:

```
from typing import List, Dict, Any
from models.data.odmantic import Purchase


class PurchaseRepository:

    def __init__(self, engine):
        self.engine = engine
```

```python
async def insert_purchase(self,
        details:Dict[str, Any]) -> bool:
    try:
        purchase = Purchase(**details)
        await self.engine.save(purchase)


    except Exception as e:
        print(e)
        return False
    return True
```

This `insert_purchase()` method shows the standard way to insert a record into the database using ODMantic. Through the engine's `save()` method, we can persist one document at a time using the model class. `AIOEngine` also provides the `save_all()` method for inserting a list of multiple documents into the associated MongoDB collection.

Now, there is no specific way to update transactions, but ODMantic allows you to fetch the record that needs to be updated. The following code can be used to update a record using ODMantic:

```python
async def update_purchase(self, id:int,
        details:Dict[str, Any]) -> bool:
    try:
        purchase = await self.engine.find_one(
            Purchase, Purchase.purchase_id == id)

        for key,value in details.items():
            setattr(purchase,key,value)

        await self.engine.save(purchase)
    except Exception as e:
        print(e)
        return False
    return True
```

After accessing and changing the field values, the fetched document object will be re-saved using the save() method to reflect the changes in physical storage. The complete process is implemented in the preceding update_purchase() transaction:

```
async def delete_purchase(self, id:int) -> bool:
    try:
        purchase = await self.engine.find_one(
            Purchase, Purchase.purchase_id == id)
        await self.engine.delete(purchase)
    except:
        return False
    return True
```

When it comes to document removal, you must fetch the document to be deleted. We pass the fetched document object to the delete() method of the engine to pursue the removal process. This implementation is shown in the delete_purchase() method.

When fetching a single document so that it can be updated or deleted, AIOEngine has a find_ one() method that requires two arguments: the model class name and the conditional expression, which involves either the id primary key or some non-unique fields. All the fields can be accessed like class variables. The following get_purchase() method retrieves a Purchase document with the specified id:

```
async def get_all_purchase(self):
    purchases = await self.engine.find(Purchase)
    return purchases


async def get_purchase(self, id:int):
    purchase = await self.engine.find_one(
        Purchase, Purchase.purchase_id == id)
    return purchase
```

The engine has a find() operation to retrieve all Purchase documents, for instance, from the database. It only needs an argument – the name of the model class. Let now apply our repository layer to the API services.

Running the CRUD transaction

For the repository classes to run, all the router services must be asynchronous. Then, we need to create the startup and shutdown event handlers for `create_db_connection()` and `close_db_connection()`, respectively, to open the connection for repository transactions. Lastly, for the repository class to work, `create_db_engine()` must be injected into each API service to derive the engine object:

```python
from fastapi import APIRouter, Depends
from fastapi.responses import JSONResponse

from models.request.purchase import PurchaseReq
from repository.odmantic.purchase import PurchaseRepository
from db_config.odmantic_config import create_db_engine,
    create_db_connection, close_db_connection

from datetime import date, datetime
from json import dumps, loads

router = APIRouter()

router.add_event_handler("startup", create_db_connection)
router.add_event_handler("shutdown", close_db_connection)

@router.post("/purchase/add")
async def add_purchase(req: PurchaseReq,
          engine=Depends(create_db_engine)):
    purchase_dict = req.dict(exclude_unset=True)
    purchase_json = dumps(purchase_dict,
              default=json_serial)
    repo:PurchaseRepository = PurchaseRepository(engine)
    result = await
          repo.insert_purchase(loads(purchase_json))
    if result == True:
       return req
    else:
       return JSONResponse(content={"message":
```

```
            "insert purchase unsuccessful"}, status_code=500)
    return req
```

At this point, we should know how to compare these mappers and drivers when it comes to the setup and procedures needed to manage MongoDB documents. Each has its strengths and weaknesses based on the code they produce and the performance, popularity, support, and complexity of its solution. Some may work on other requirements, while others may not. The final ODM we will cover focuses on being the lightest and least obtrusive mapper. It aims to fit into an existing application without generating syntax and performance problems.

Creating CRUD transactions using MongoFrames

If you are tired of using complicated and heavy-loaded ODMs, then MongoFrames is ideal for your requirements. MongoFrames is one of the newest ODMs and is very convenient to use, especially when building a new repository layer for an already existing complex and legacy FastAPI microservice application. But this mapper can only create synchronous and standard types of CRUD transactions.

But before we proceed, let us install the extension module using `pip`:

```
pip install MongoFrames
```

Creating the database connection

The MongoFrames platform runs on top of PyMongo, which is why it cannot build an asynchronous repository layer. To create the database connection, it uses the `MongoClient` API class from the `pymongo` module, with the database URL in string format. Unlike in the other ODMs, where we create a client variable, in this mapper, we access the `variable _client` class from the `Frame` API class to refer to the client connection object. The following code shows `create_db_client()`, which will open the database connection for our app, and `disconnect_db_client()`, which will close this connection:

```
from pymongo import MongoClient
from mongoframes import Frame


def create_db_client():
    Frame._client =
        MongoClient('mongodb://localhost:27017/obrs')


def disconnect_db_client():
    Frame._client.close()
```

Just like in the previous ODMs, we need event handlers to execute these core methods to start building the model and repository layers.

Building the model layer

The process of creating model classes in MongoFrames is called **framing** because it uses the `Frame` API class to define the model classes. Once inherited, `Frame` does not require a model class to define its attributes. It uses the `_fields` property to contain all the necessary fields of the document without indicating any metadata. The following model classes are defined by the `Frame` API class:

```python
from mongoframes import Frame, SubFrame


class Book(Frame):
    _fields = {
        'id ',
        'isbn',
        'author',
        'date_published',
        'title',
        'edition',
        'price',
        'category'
    }
    _collection = "book"


class Category(SubFrame):

    _fields = {
        'id',
        'name',
        'description',
        'date_added'
        }

    _collection = "category"

class Reference(Frame):
```

```
    _fields = {
        'id',
        'name',
        'description',
        'categories'
        }


    _collection = "reference"
```

A `Frame` model class can wrap a document in dictionary form or in a `kwargs` that contains the key-value details of the document's structure. It can also provide attributes and helper methods that can help pursue CRUD transactions. All the fields of the model class can be accessed through dot (`.`) notation, just like typical class variables.

Creating the document association

We need to define the `SubFrame` model before creating associations among these documents. A `SubFrame` model class is mapped to an embedded document structure and has no collection table of its own. The MongoFrames mapper provides operations that allow you to append, update, remove, and query the `SubFrame` class of the `Frame` instance. These operations will determine the type of association among documents since the field references of `Frame` do not have specific field types. The `Reference` document, for instance, will have a list of categories linked to its `categories` field because our transaction will build that association as designed. A `Book` document, on the other hand, will refer to a `Category` child document through its `category` field because a transaction will build that association at runtime. So, MongoFrames is both restrained and non-strict when it comes to defining the type of association among these documents.

Creating the repository layer

The `Frame` API class provides the model classes and the necessary helper methods to implement the asynchronous repository transactions. The following code shows an implementation of a repository class that uses MongoFrames to create its CRUD transactions:

```python
from mongoframes.factory.makers import Q
from models.data.mongoframe import Book, Category
from typing import List, Dict, Any


class BookRepository:
    def insert_book(self,
            details:Dict[str, Any]) -> bool:
```

```
        try:
            book = Book(**details)
            book.insert()


        except Exception as e:
            return False
        return True
```

The given `insert_book()` transaction inserts a book instance into its mapped collection. The Frame API provides an `insert()` method that saves the given model object into the database. It also has `insert_many()`, which inserts a list of multiple BSON documents or a list of model instances. The following script shows how to create an *UPDATE* transaction in MongoFrames:

```
def update_book(self, id:int,
        details:Dict[str, Any]) -> bool:
    try:
        book = Book.one(Q.id == id)
        for key,value in details.items():
            setattr(book,key,value)
        book.update()


    except:
        return False
    return True
```

The given `update_book()` transaction shows that the `Frame` model class also has an `update()` method, which recognizes and saves the changes reflected in the field values of a document object right after fetching them from the collection. A similar process is applied to the `delete_book()` process, which calls the `delete()` operation of the document object right after fetching it from the collection:

```
def delete_book(self, id:int) -> bool:
    try:
        book = Book.one(Q.id == id)
        book.delete()
    except:
        return False
    return True
```

When creating query transactions, the Frame API provides two class methods – the many() method, which extracts all BSON documents, and the one() method, which returns a single document object. Both operations can accept a query expression as an argument if there are any constraints. Moreover, MongoFrames has a Q query maker class that's used to build conditionals in a query expression. The expression starts with Q, followed by dot (.) notation to define the field name or path – for example, Q.categories.fiction – followed by an operator (for example, ==, !=, >, >=, <, or <=) and finally a value. The following code shows examples of the query transactions being translated using the MongoFrames ODM syntax:

```
def get_all_book(self):
    books = [b.to_json_type() for b in Book.many()]
    return books


def get_book(self, id:int):
    book = Book.one(Q.id == id).to_json_type()
    return book
```

The get_book() method shows how to extract a single Book document with a Q expression that filters the correct id, while get_all_book() retrieves all Book documents without any constraints.

The many() operator returns a list of Frame objects, while the one() operator returns a single Frame instance. To convert the result into JSON-able components, we need to invoke the to_json_type() method in each Frame instance.

As explained earlier, adding embedded documents is determined by the operation and not by the model attributes. In the following add_category() transaction, it is clear that a Category object has been assigned to a category field of a Book instance, even if the field is not defined to refer to an embedded document of the Category type. Instead of throwing an exception, MongoFrame will update the Book document right after the update() call:

```
def add_category(self, id:int,
            category:Category) -> bool:
    try:
        book = Book.one(Q.id == id)
        book.category = category
        book.update()
    except:
        return False
    return True
```

Now, it is time to apply these CRUD transactions to our API services.

Applying the repository layer

Our repository classes will not work if we do not inject the `create_db_client()` injectable into the router. The following solution injects the component into `APIRouter`, even if it is acceptable to inject it into each API service implementation:

```python
from fastapi import APIRouter, Depends
from fastapi.responses import JSONResponse

from models.request.category import BookReq
from repository.mongoframe.book import BookRepository
from db_config.mongoframe_config import create_db_client

from datetime import date, datetime
from json import dumps, loads

router = APIRouter(
        dependencies=[Depends(create_db_client)])


@router.post("/book/create")
def create_book(req:BookReq):
    book_dict = req.dict(exclude_unset=True)
    book_json = dumps(book_dict, default=json_serial)
    repo:BookRepository = BookRepository()
    result = repo.insert_book(loads(book_json))
    if result == True:
        return req
    else:
        return JSONResponse(content={"message":
            "insert book unsuccessful"}, status_code=500)
```

The `create_book()` service uses `BookRepository` to insert book details into the MongoDB database. In general, MongoFrames has an easy setup because it requires fewer configuration details for creating the database connection, building the model layer, and implementing the repository transactions. Its platform can be adapted to the existing requirements of the application and can easily reflect changes if modifications need to be made to its mapping mechanisms.

Summary

In this chapter, we looked at various ways to manage data using MongoDB. We utilized MongoDB to store non-relational data for our *online book reselling system* since we expect the data to become large when information is exchanged between the book buyers and resellers. Additionally, the details involved in the transactions are mainly strings, floats, and integers, which are all order and purchase values that will be easier to mine and analyze if they're stored in schema-less storage.

This chapter took the non-relational data management roadmap for utilizing the data in sales forecasting, regression analysis of book readers' demands, and other descriptive data analysis forms.

First, you learned how the PyMongo and Motor drivers connect the FastAPI application to the MongoDB database. After understanding the nuts and bolts of creating CRUD transactions using these drivers, you learned that ODM is the better option for pursuing MongoDB connectivity. We explored the features of MongoEngine, Beanie, ODMantic, and MongoFrames and studied their strengths and weaknesses as ODM mappers. All these ODMs can be integrated well with the FastAPI platform and provide the application with a standardized way to back up data.

Now that we've spent two chapters covering data management, in the next chapter, we will learn how to secure our FastAPI microservice applications.

7
Securing the REST APIs

Building microservices means exposing the entire application to the worldwide web. For every request-response transaction, the client accesses the endpoint of the API publicly, which poses potential risks to the application. Unlike web-based applications, API services have weak mechanisms to manage user access using login controls. Thus, this chapter will provide several ways to protect the API services created using the FastAPI framework.

There is no such thing as perfect security. The main goal is to establish policies and solutions related to the *confidentiality*, *integrity*, and *availability* of these services. The *confidentiality policy* requires tokens, encryption and decryption, and certificates as mechanisms to make some APIs private. On the other hand, the *integrity policy* involves maintaining the data exchange as authentic, accurate, and reliable by using a "state" and hashed codes during the authentication and authorization process. The *availability policy* means protecting the endpoint access from DoS attacks, phishing, and timing attacks using reliable tools and Python modules. Overall, these three aspects of the security model are the essential elements to consider when building security solutions for our microservices.

Although FastAPI has no built-in security framework, it supports different authentication modes such as *Basic* and *Digest*. It also has built-in modules that implement security specifications such as *OAuth2*, *OpenID*, and *OpenAPI*. The following main topics will be covered in this chapter to explain and illustrate the concepts and solutions for securing our FastAPI services:

- Implementing Basic and Digest authentication
- Implementing password-based authentication
- Applying JWTs
- Creating scope-based authorization
- Building the authorization code flow
- Applying the OpenID Connect specification
- Using built-in middleware for authentication

Technical requirements

The software prototype for this chapter is a *secure online auction system* designed to manage online bidding on various items auctioned by its registered users. The system can bid on any items based within a price range and even declare those who won the bidding. The system needs to secure some sensitive transactions to avoid data breaches and biased results. The prototype will be using *SQLAlchemy* as the ORM for managing data. There will be 10 versions of our prototype and each will showcase a different authentication scheme. All 10 of these projects (ch07a to ch07j) can be found here: https://github.com/PacktPublishing/Building-Python-Microservices-with-FastAPI.

Implementing Basic and Digest authentication

The Basic and Digest authentication schemes are the easiest authentication solutions that we can use to secure API endpoints. Both schemes are alternative authentication mechanisms that can be applied to small and low-risk applications without requiring complex configuration and coding. Let us now use these schemes to secure our prototype.

Using Basic authentication

The most straightforward way to secure the API endpoint is the *Basic authentication* approach. However, this authentication mechanism must not be applied to high-risk applications because the credentials, commonly a username and password, sent from the client to the security scheme provider are in the *Base64-encoded* format, which is vulnerable to many attacks such as *brute force*, *timing attacks*, and *sniffing*. Base64 is not an encryption algorithm but simply a way of representing the credentials in *ciphertext* format.

Applying HttpBasic and HttpBasicCredentials

The prototype, ch07a, uses the Basic authentication mode to secure its administration and bidding and auctioning transactions. Its implementation in the /security/secure.py module is shown in the following code:

```
from passlib.context import CryptContext
from fastapi.security import HTTPBasicCredentials
from fastapi.security import HTTPBasic

from secrets import compare_digest
from models.data.sqlalchemy_models import Login

crypt_context = CryptContext(schemes=["sha256_crypt",
                        "md5_crypt"])
```

```
http_basic = HTTPBasic()
```

The FastAPI framework supports different authentication modes and specifications through its fastapi.security module. To pursue the *Basic* authentication scheme, we need to instantiate the HTTPBasic class of the module and inject it into each API service to secure the endpoint access. The http_basic instance, once injected into the API services, causes the browser to pop up a login form, through which we type the username and password credentials. Logging in will trigger the browser to send a header with the credentials to the application. If the application encounters a problem with receiving it, the HTTPBasic scheme will throw an *HTTP status code 401* with an *"Unauthorized"* message. If there are no errors in the form handling, the application must receive a WWW-Authenticate header with a Basic value and an optional realm parameter.

On the other hand, the /ch07/login service will call the authentication() method to verify whether the browser credentials are authentic and correct. We need to be very careful in accepting user credentials from browsers since they are prone to various attacks. First, we can require endpoint users to use an *email address* as their username and require long passwords with a combination of different characters, numbers, and symbols. All stored passwords must be encoded using the most reliable encryption tools, such as the CryptContext class from the passlib module. The passlib extension provides more secured hashing algorithms than any Python encryption module. Our application uses SHA256 and MD5 hashing algorithms instead of the recommended bcrypt, which is slower and prone to attacks.

Second, we can avoid storing the credentials in the source code and use database storage or a .env file instead. The authenticate() method checks the credentials against the Login database record provided by the API service for correctness.

Lastly, always use the compare_digest() from the secret module when comparing credentials from the browser with the Login credentials stored in the database. This function randomly compares two strings while guarding the operation against timing attacks. A *timing attack* is a kind of attack that compromises the crypto-algorithm execution, which happens when there is a linear comparison of strings in the system:

```
def verify_password(plain_password, hashed_password):
    return crypt_context.verify(plain_password,
        hashed_password)


def authenticate(credentials: HTTPBasicCredentials,
        account:Login):
    try:
        is_username = compare_digest(credentials.username,
            account.username)
```

```
        is_password = compare_digest(credentials.password,
            account.username)
        verified_password =
            verify_password(credentials.password,
                account.passphrase)
        return (verified_password and is_username and
            is_password)
    except Exception as e:
        return False
```

Our `authenticate()` method has all the needed requirements to help reduce attacks from outside factors. But the ultimate solution to secure Basic authentication is to install and configure a *Transport Layer Security* (*TLS*) (or *HTTPS*, or *SSL*) connection for the application.

Now, we need to implement a `/ch07/login` endpoint to apply the *Basic* authentication scheme. The `http_basic` instance is injected into this API service to extract `HTTPBasicCredentials`, which is the object that contains the *username* and *password* details from the browser. This service is also the one that calls the `authenticate()` method to check the user credentials. If the method returns a `False` value, the service will raise an *HTTP status code 400* with an *"Incorrect credentials"* message:

```
from fastapi import APIRouter, Depends, HTTPException

from fastapi.security import HTTPBasicCredentials
from security.secure import authenticate,
        get_password_hash, http_basic


router = APIRouter()


@router.get("/login")
def login(credentials: HTTPBasicCredentials =
    Depends(http_basic), sess:Session = Depends(sess_db)):

    loginrepo = LoginRepository(sess)
    account = loginrepo.get_all_login_username(
                    credentials.username)
    if authenticate(credentials, account) and
            not account == None:
        return account
```

```
    else:
        raise HTTPException(
            status_code=400,
            detail="Incorrect credentials")

@router.get("/login/users/list")
def list_all_login(credentials: HTTPBasicCredentials =
    Depends(http_basic), sess:Session = Depends(sess_db)):
    loginrepo = LoginRepository(sess)
    users = loginrepo.get_all_login()
    return jsonable_encoder(users)
```

Each endpoint of the *online auction system* must have the injected `http_basic` instance to secure it from public access. For instance, the cited `list_all_login()` service can only return a list of all users if the user is an authenticated one. By the way, there is no reliable procedure to log off using *Basic* authentication. If the `WWW-Authenticate` header has been issued and recognized by the browser, we will seldom see the login form of the browser pop up.

Executing the login transaction

We can use either the `curl` command or the browser to perform the `/ch07/login` transaction. But to highlight the support of FastAPI, we will be using its OpenAPI dashboard to run `/ch07/login`. After accessing `http://localhost:8000/docs` on the browser, locate the `/ch07/login` GET transaction and click the **Try it out** button. The browser's login form, as shown in *Figure 7.1*, will pop up after clicking the button:

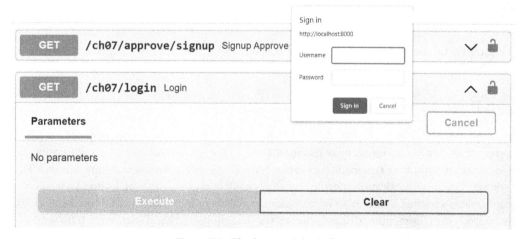

Figure 7.1 – The browser's login form

After the **Username** and **Password** input, click the **Sign in** button on the login form to check whether the credentials are in the database. Otherwise, the app has `/ch07/signup/add` and `/ch07/approve/signup` to add the user credentials you want to test. Remember that all stored passwords are encrypted. *Figure 7.2* shows how `/ch07/login` will output the user's `Login` record after the authentication process finds that the user credentials are valid:

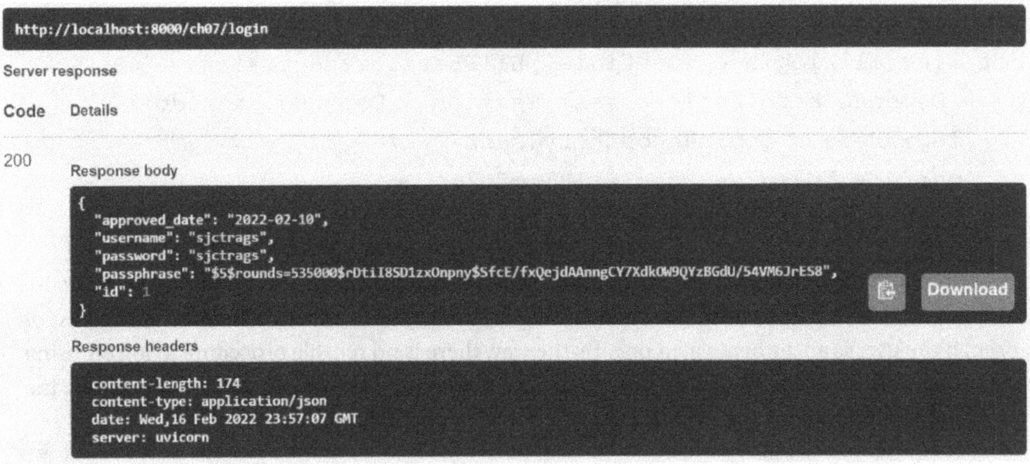

```
http://localhost:8000/ch07/login
```

Server response

Code   Details

200
       Response body
```
{
    "approved_date": "2022-02-10",
    "username": "sjctrags",
    "password": "sjctrags",
    "passphrase": "$5$rounds=535000$rDtiI8SD1zxOnpny$SfcE/fxQejdAAnngCY7XdkOW9QYzBGdU/54VM6JrES8",
    "id": 1
}
```
                                                                          Download

       Response headers
```
content-length: 174
content-type: application/json
date: Wed,16 Feb 2022 23:57:07 GMT
server: uvicorn
```

Figure 7.2 – The /login response

Now that the user is authenticated, run `/ch07/login/users/list` through the OpenAPI dashboard to retrieve the list of login details. The `uvicorn` server log will show the following log message:

```
INFO: 127.0.0.1:53150 - "GET /ch07/login/users/list HTTP/1.1"
200 OK
```

This means that the user is authorized to run the endpoint. Now, let us apply the Digest authentication scheme to our prototype.

Using Digest authentication

Digest authentication is more secure than the Basic scheme because the former needs to hash the user credentials first before sending the hashed version to the application. Digest authentication in FastAPI does not include an automatic encryption process of user credentials using the default *MD5* encryption. It is an authentication scheme that stores credentials in a `.env` or `.config` property file and creates a hashed string value for these credentials before the authentication. The `ch07b` project applies the Digest authentication scheme to secure the bidding and auctioning transactions.

Generating the hashed credentials

So, before we start the implementation, we first need to create a custom utility script, `generate_hash.py`, that generates a digest in binary form using Base64 encoding. The script must have the following code:

```
from base64 import urlsafe_b64encode
h = urlsafe_b64encode(b"sjctrags:sjctrags")
```

The `urlsafe_b64encode()` function from the `base64` module creates a digest in binary format from the `username:password` credential format. After running the script, we save the digest value anywhere safe, but not in the source code.

Passing the user credentials

Aside from the digest, we also need to save the user credentials for the Digest scheme provider later. Unlike the standard Digest authentication procedure, where the user negotiates with the browser, FastAPI requires storing the user credentials in a `.env` or `.config` file inside our application to be retrieved by the authentication process. In the `ch07b` project, we save the username and password inside the `.config` file, in this manner:

```
[CREDENTIALS]
USERNAME=sjctrags
PASSWORD=sjctrags
```

Then, we create a parser through the `ConfigParser` utility to extract the following details from the `.config` file and build a `dict` out of the serialized user details. The following `build_map()` is an example of the parser implementation:

```
import os
from configparser import ConfigParser


def build_map():
    env = os.getenv("ENV", ".config")
    if env == ".config":
        config = ConfigParser()
        config.read(".config")
        config = config["CREDENTIALS"]
    else:
        config = {
            "USERNAME": os.getenv("USERNAME", "guest"),
```

```
                "PASSWORD": os.getenv("PASSWORD", "guest"),
        }


    return config
```

Using HTTPDigest and HTTPAuthorizationCredentials

The FastAPI framework has an `HTTPDigest` from its `fastapi.security` module that implements a Digest authentication scheme with a different approach to managing user credentials and generating the digest. Unlike in Basic authentication, the `HTTPDigest` authentication process happens at the `APIRouter` level. We inject the following `authenticate()` dependable into the API services through the HTTP operator, including `/login`, where the authentication starts:

```
from fastapi import Security, HTTPException, status
from fastapi.security import HTTPAuthorizationCredentials
from fastapi.security import HTTPDigest
from secrets import compare_digest
from base64 import standard_b64encode


http_digest = HTTPDigest()


def authenticate(credentials:
    HTTPAuthorizationCredentials = Security(http_digest)):

    hashed_credentials = credentials.credentials
    config = build_map()
    expected_credentials = standard_b64encode(
        bytes(f"{config['USERNAME']}:{config['PASSWORD']}",
            encoding="UTF-8")
    )
    is_credentials = compare_digest(
            bytes(hashed_credentials, encoding="UTF-8"),
                expected_credentials)

    if not is_credentials:
        raise HTTPException(
            status_code=status.HTTP_401_UNAUTHORIZED,
            detail="Incorrect digest token",
```

```
        headers={"WWW-Authenticate": "Digest"},
    )
```

The `authenticate()` method is where the `http_digest` is injected to extract the `HTTPAuthorizationCredentials` that contains the digest byte value. After extraction, it checks whether the digest matches the credentials saved in the `.config` file. We also use `compare_digest` to compare `hashed_credentials` from the header and the Base64-encoded credentials from the `.config` file.

Executing the login transaction

After implementing the `authenticate()` method, we inject it into the API services, not in the method parameter, but in its HTTP operator. Notice that the `http_digest` object is not injected directly into the API services, unlike in the *Basic* authentication scheme. The following implementation shows how the `authenticate()` dependable is applied to secure all the crucial endpoints of the application:

```
from security.secure import authenticate


@router.get("/login", dependencies=[Depends(authenticate)])
def login(sess:Session = Depends(sess_db)):
    return {"success": "true"}


@router.get("/login/users/list",
        dependencies=[Depends(authenticate)])
def list_all_login(sess:Session = Depends(sess_db)):
    loginrepo = LoginRepository(sess)
    users = loginrepo.get_all_login()
    return jsonable_encoder(users)
```

Since the Digest authentication scheme behaves like the *OpenID authentication*, we will be using the `curl` command to run `/ch07/login`. The crucial part of the command is the issuance of the `Authorization` header with the value containing the Base64-encoded `username:password` digest generated by the `generate_hash.py` script we executed beforehand. The following `curl` command is the correct way of logging into our FastAPI application that uses the Digest authentication scheme:

```
curl --request GET --url http://localhost:8000/ch07/login
--header "accept: application/json"                    --header
"Authorization: Digest c2pjdHJhZ3M6c2pjdHJhZ3M=" --header
"Content-Type: application/json"
```

We also use the same command to run the rest of the secured API services.

Most enterprise applications nowadays seldom use Basic and Digest authentication schemes because of their vulnerability to many attacks. More than that, both authentication schemes require sending credentials to the secured API services, which is also another risk. Moreover, at the time of writing, FastAPI does not yet fully support the standard Digest authentication, which is also a disadvantage to other applications that need the standard one. So, let us now explore the solution to secure API endpoints using the *OAuth 2.0 specification*.

Implementing password-based authentication

The *OAuth 2.0 specification*, or OAuth2, is the most preferred solution for authenticating API endpoint access. The OAuth2 authorization framework defines the four authorization flows, which are *implicit, client credentials, authorization code*, and *resource password flows*. The first three of these can be used with third-party authentication providers, which will authorize the access of the API endpoints. In the FastAPI platform, the resource password flow can be customized and implemented within the application to carry out the authentication procedure. Let us now explore how FastAPI supports the OAuth2 specification.

Installing the python-multipart module

Since OAuth2 authentication will not be possible without a form handling procedure, we need to install the python-multipart module before pursuing the implementation part. We can run the following command to install the extension:

```
pip install python-multipart
```

Using OAuth2PasswordBearer and OAuth2PasswordRequestForm

The FastAPI framework fully supports OAuth2, especially the password flow type of the OAuth2 specification. Its fastapi.security module has an OAuth2PasswordBearer that serves as the provider for password-based authentication. It also has OAuth2PasswordRequestForm, which can declare a form body with required parameters, username and password, and some optional ones such as scope, grant_type, client_id, and client_secret. This class is directly injected into the /ch07/login API endpoint to extract all the parameter values from the browser's login form. But it is always an option to use Form(...) to capture all the individual parameters.

So, let us start the solution by creating the OAuth2PasswordBearer to be injected into a custom function dependency that will validate the user credentials. The following implementation shows that get_current_user() is the injectable function in our new application, ch07c, which utilizes the oath2_scheme injectable to extract a token:

```
from fastapi.security import OAuth2PasswordBearer
from sqlalchemy.orm import Session
from repository.login import LoginRepository
```

```
from db_config.sqlalchemy_connect import sess_db
oauth2_scheme =
    OAuth2PasswordBearer(tokenUrl="ch07/login/token")


def get_current_user(token: str = Depends(oauth2_scheme),
            sess:Session = Depends(sess_db) ):
    loginrepo = LoginRepository(sess)
    user = loginrepo.get_all_login_username(token)
    if user == None:
        raise HTTPException(
            status_code=status.HTTP_401_UNAUTHORIZED,
            detail="Invalid authentication credentials",
            headers={"WWW-Authenticate": "Bearer"},
        )
    return user
```

For the resource password flow, injecting `oauth2_scheme` will return a `username` as a token. `get_current_user()` will check whether that username belongs to a valid user account stored in the database.

Executing the login transaction

In this authentication scheme, `/ch07/login/token` is also the `tokenUrl` parameter of `OAuth2PasswordBearer`. The `tokenUrl` parameter is required for password-based OAuth2 authentication because this is the endpoint service that will capture the user credentials from the browser's login form. `OAuth2PasswordRequestForm` is injected into `/cho07/login/token` to retrieve the `username`, `password`, and `grant_type` parameters of the unauthenticated user. These three parameters are the essential requirements to invoke `/ch07/login/token` for *token* generation. This dependency is shown in the following implementation of the login API service:

```
from sqlalchemy.orm import Session
from db_config.sqlalchemy_connect import sess_db
from repository.login import LoginRepository
from fastapi.security import OAuth2PasswordRequestForm
from security.secure import get_current_user, authenticate


@router.post("/login/token")
def login(form_data: OAuth2PasswordRequestForm = Depends(),
            sess:Session = Depends(sess_db)):
```

```
username = form_data.username
password = form_data.password
loginrepo = LoginRepository(sess)
account = loginrepo.get_all_login_username(username)
if authenticate(username, password, account) and
        not account == None:
    return {"access_token": form_data.username,
            "token_type": "bearer"}
else:
    raise HTTPException(
        status_code=400,
            detail="Incorrect username or password")
```

Aside from verifying from the database, the login() service will also check whether the password value matches the encrypted passphrase from the queried account. If all the verification succeeds, /ch07/login/token must return a JSON object with the required properties, access_token and token_type. The access_token property must have the username value, and token_type the "bearer" value.

Instead of creating a custom frontend for the login form, we will be utilizing the OAuth2 form provided by OpenAPI in the framework. We just click the **Authorize** button on the upper-right-hand side of the OpenAPI dashboard, as shown in *Figure 7.3*:

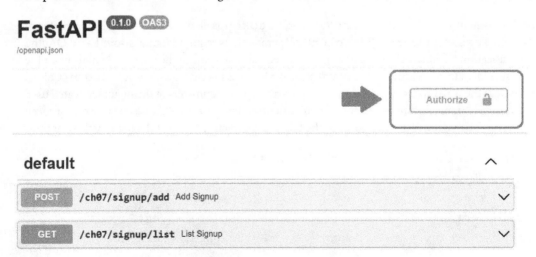

Figure 7.3 – The Authorize button

The button will trigger a built-in login form to pop up, shown in *Figure 7.4*, which we can use to test our solution:

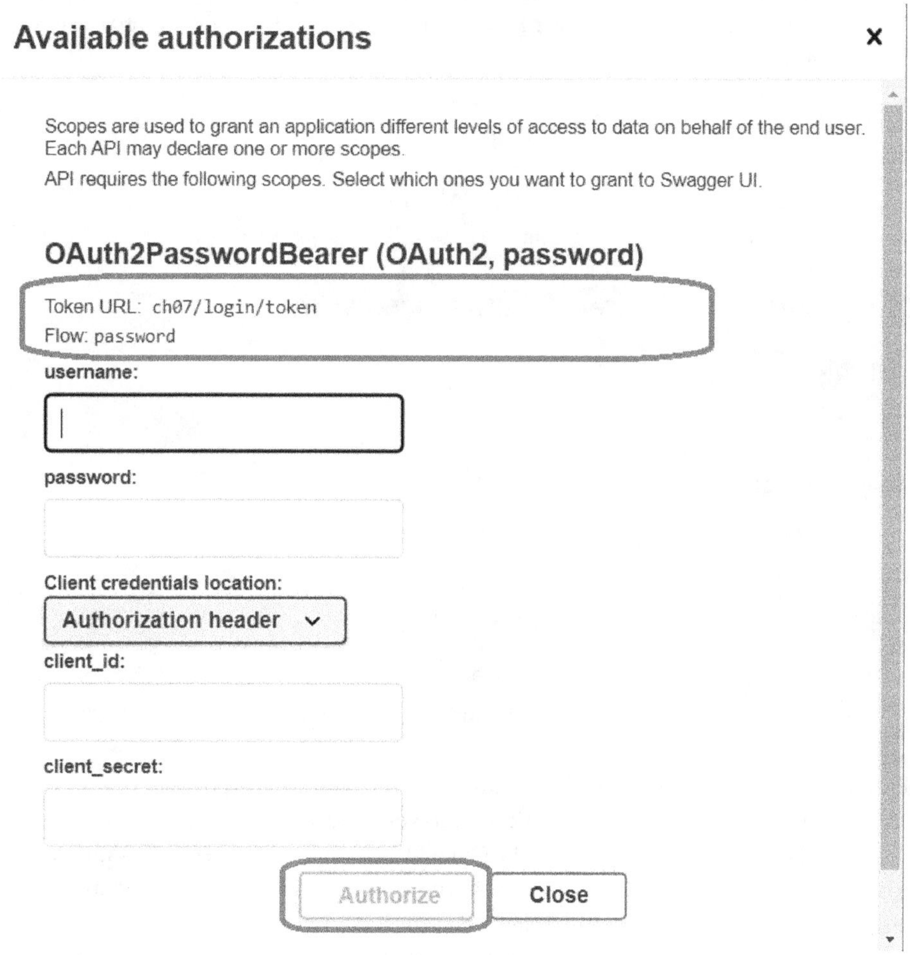

Available authorizations   ✕

Scopes are used to grant an application different levels of access to data on behalf of the end user. Each API may declare one or more scopes.

API requires the following scopes. Select which ones you want to grant to Swagger UI.

OAuth2PasswordBearer (OAuth2, password)

Token URL: `ch07/login/token`
Flow: `password`

username:

password:

Client credentials location:

    Authorization header   ⌄

client_id:

client_secret:

    Authorize   Close

Figure 7.4 – The OAuth2 login form

Everything is fine if the OAuth2 login form detects the correct `tokenURL` specified in the `OAuth2PasswordBearer` instantiation. The OAuth2 flow or `grant_type` indicated in the login form must be `"password"`. After logging the verified credential, the form's **Authorize** button will redirect the user to an authorization form, shown in *Figure 7.5*, which will prompt the user to log out or proceed with the authenticated access:

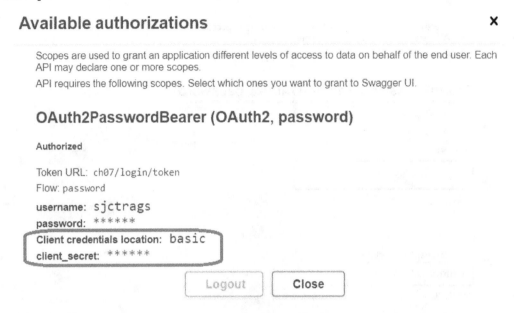

Figure 7.5 – The authorization form

Generally, the OAuth2 specification recognizes two client or application types: *confidential* and *public* clients. The confidential clients utilize authentication servers for security, such as in this *online auction system* that uses the FastAPI server through the OpenAPI platform. In its setup, it is not mandatory to provide the `client_id` and `client_secret` values to the login form since the server will generate these parameters during the authentication process. But unfortunately, these values are not revealed to the client, as shown in *Figure 7.5*. On the other hand, the public clients do not have any means to generate and use client secrets as in typical web-based and mobile applications. Therefore, these applications must include `client_id`, `client_secret`, and other required parameters during login.

Securing the endpoints

To secure the API endpoints, we need to inject the `get_current_user()` method into each API service method. The following is an implementation of a secured `add_auction()` service that utilizes the `get_current_user()` method:

```
@router.post("/auctions/add")
def add_auction(req: AuctionsReq,
```

```
    current_user: Login = Depends(get_current_user),
    sess:Session = Depends(sess_db)):
  auc_dict = req.dict(exclude_unset=True)
  repo:AuctionsRepository = AuctionsRepository(sess)
  auction = Auctions(**auc_dict)
  result = repo.insert_auction(auction)
  if result == True:
      return auction
  else:
      return JSONResponse(content=
        {'message':'create auction problem encountered'},
          status_code=500)
```

The get_current_user() injectable will return a valid Login account if the access is allowed. Moreover, you will notice that all padlock icons of the secured API endpoints that include /ch07/auctions/add, shown in *Figure 7.6*, are closed. This indicates that they are ready to be executed since the user is already an authenticated one:

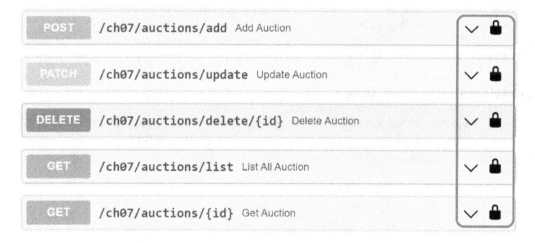

Figure 7.6 – An OpenAPI dashboard showing secured APIs

This solution is a problem for an open network setup, for instance, because the token used is a password. This setup allows attackers to easily forge or modify the token during its transmission from the issuer to the client. One way to protect the token is to use a **JSON Web Token (JWT)**.

Applying JWTs

JWT is an open source standard used to define a solution for sending any information during the authentication and authorization between issuers and clients. Its goal is to generate `access_token` properties that are digitally signed, URL-safe, and always verifiable by the client. However, it is not perfectly safe because anyone can decode the token if needed. Thus, it is advisable not to include all the valuable and confidential information in the token string. A JWT is an effective way of providing OAuth2 and OpenID specifications with more reliable tokens than passwords.

Generating the secret key

But before we start building the authentication scheme, we first need to generate a *secret key*, which is an essential element in creating the *signature*. The JWT has a **JSON Object Signing and Encryption (JOSE)** *header*, which is the metadata that describes which algorithm to use for plain-text encoding, while the *payload* is the data we need to encode into the token. When the client requests to log in, the authorization server signs a JWT using a *signature*. But the signature will only be generated by the algorithm indicated in the header, which will take the header, payload, and secret key as inputs. This *secret key* is a Base64-encoded string manually created outside of the server and should be stored separately within the authorization server. `ssh` or `openssl` is the appropriate utility to generate this long and randomized key. Here, in `ch07d`, we run the following `openssl` command from a GIT tool or any SSL generator to create the key:

```
openssl rand -hex 32
```

Creating the access_token

In the `ch07d` project, we will store the *secret key* and *algorithm type* in some reference variables in its `/security/secure.py` module script. These variables are used by the JWT-encoding procedure to generate the token, as shown in the following code:

```
from jose import jwt, JWTError
from datetime import datetime, timedelta


SECRET_KEY = "tbWivbkVxfsuTxCP8A+Xg67LcmjXXl/sszHXwH+TX9w="
ALGORITHM = "HS256"
ACCESS_TOKEN_EXPIRE_MINUTES = 30


def create_access_token(data: dict,
            expires_after: timedelta):
    plain_text = data.copy()
    expire = datetime.utcnow() + expires_after
```

```
plain_text.update({"exp": expire})
encoded_jwt = jwt.encode(plain_text, SECRET_KEY,
        algorithm=ALGORITHM)
return encoded_jwt
```

Within the JWT Python extension, we chose the `python-jose` module to generate the token because it is reliable and has additional cryptographic functions that can sign complex data content. Install this module first using the `pip` command before using it.

So now, the `/ch07/login/token` endpoint will invoke the `create_access_token()` method to request the JWT. The *login* service will provide the data, usually `username`, to comprise the payload portion of the token. Since the JWT must be short-lived, the process must update the `expire` portion of the payload to some `datetime` value in minutes or seconds suited to the application.

Creating the login transaction

The implementation of the *login* service is similar to the previous password-based OAuth2 authentication, except that this version has a `create_access_token()` call for the JWT generation to replace the password credential. The following script shows the `/ch07/login/token` service of the ch07d project:

```
@router.post("/login/token")
def login(form_data: OAuth2PasswordRequestForm = Depends(),
        sess:Session = Depends(sess_db)):
    username = form_data.username
    password = form_data.password
    loginrepo = LoginRepository(sess)
    account = loginrepo.get_all_login_username(username)
    if authenticate(username, password, account):
        access_token = create_access_token(
          data={"sub": username},
           expires_after=timedelta(
             minutes=ACCESS_TOKEN_EXPIRE_MINUTES))
        return {"access_token": access_token,
            "token_type": "bearer"}
    else:
        raise HTTPException(
            status_code=400,
            detail="Incorrect username or password")
```

The endpoint should still return `access_token` and `token_type` since this is still a password-based OAuth2 authentication, which retrieves the user credentials from `OAuth2PasswordRequestForm`.

Accessing the secured endpoints

As with the previous OAuth2 schemes, we need to inject `get_current_user()` into every API service to impose security and restrict access. The injected `OAuthPasswordBearer` instance will return the JWT for payload extraction using the JOSE decoders with the specified decoding algorithm. If the token is tampered with, modified, or expired, the method will throw an change to - exception. Otherwise, we need to continue the payload data extraction, retrieve the username, and store that in an `@dataclass` instance, such as `TokenData`. Then, the username will undergo further verification, such as checking the database for a `Login` account with that username. The following snippet shows this decoding process, found in the `/security/secure.py` module of the `ch07d` project:

```
from models.request.tokens import TokenData
from fastapi.security import OAuth2PasswordBearer
from jose import jwt, JWTError

from models.data.sqlalchemy_models import Login
from sqlalchemy.orm import Session
from db_config.sqlalchemy_connect import sess_db
from repository.login import LoginRepository

from datetime import datetime, timedelta

oauth2_scheme =
    OAuth2PasswordBearer(tokenUrl="ch07/login/token")

def get_current_user(token: str = Depends(oauth2_scheme),
    sess:Session = Depends(sess_db)):
    credentials_exception = HTTPException(
        status_code=status.HTTP_401_UNAUTHORIZED,
        detail="Could not validate credentials",
        headers={"WWW-Authenticate": "Bearer"}
    )
    try:
        payload = jwt.decode(token, SECRET_KEY,
            algorithms=[ALGORITHM])
```

```
            username: str = payload.get("sub")
        if username is None:
            raise credentials_exception
        token_data = TokenData(username=username)
    except JWTError:
        raise credentials_exception

    loginrepo = LoginRepository(sess)
    user =
      loginrepo.get_all_login_username(token_data.username)
    if user is None:
        raise credentials_exception
    return user
```

get_current_user() must be injected into each service implementation to restrict access from users. But this time, the method will not only verify the credentials but also perform *JWT payload decoding*. The next step is adding *user authorization* to the OAuth2 solution.

Creating scope-based authorization

FastAPI fully supports *scope-based authentication*, which uses the scopes parameter of the OAuth2 protocol to specify which endpoints are accessible to a group of users. A scopes parameter is a kind of permission placed in a token to provide additional fine-grained restrictions to users. In this version of the project, ch07e, we will be showcasing OAuth2 password-based authentication with user authorization.

Customizing the OAuth2 class

First, we need to create a custom class that inherits the properties of the OAuth2 API class from the fastapi.security module to include the scopes parameter or "role" options in the user credentials. The following is the OAuth2PasswordBearerScopes class, a custom OAuth2 class that will implement the authentication flow with authorization:

```
class OAuth2PasswordBearerScopes(OAuth2):
    def __init__(
        self,
        tokenUrl: str,
        scheme_name: str = None,
        scopes: dict = None,
        auto_error: bool = True,
```

```
):
if not scopes:
    scopes = {}
flows = OAuthFlowsModel(
    password={"tokenUrl": tokenUrl, "scopes": scopes})
super().__init__(flows=flows,
    scheme_name=scheme_name, auto_error=auto_error)

async def __call__(self, request: Request) ->
        Optional[str]:
    header_authorization: str =
        request.headers.get("Authorization")
    ... ... ... ... ... ...
    return param
```

This OAuth2PasswordBearerScopes class requires two constructor parameters, tokenUrl and scopes, to pursue an auth flow. OAuthFlowsModel defines the scopes parameter as part of the user credentials for authentication using the Authorization header.

Building the permission dictionary

Before we proceed with the auth implementation, we need to first build the scopes parameters that the OAuth2 scheme will be applying during authentication. This setup is part of the OAuth2PasswordBearerScopes instantiation, where we assign these parameters to its scopes parameter. The following script shows how all the custom-defined user scopes are saved in a *dictionary*, with the *keys* as the scope names and the *values* as their corresponding descriptions:

```
oauth2_scheme = OAuth2PasswordBearerScopes(
    tokenUrl="/ch07/login/token",
    scopes={"admin_read":
            "admin role that has read only role",
        "admin_write":
            "admin role that has write only role",
        "bidder_read":
            "customer role that has read only role",
        "bidder_write":
            "customer role that has write only role",
        "auction_read":
```

```
            "buyer role that has read only role",
        "auction_write":
            "buyer role that has write only role",
        "user":"valid user of the application",
        "guest":"visitor of the site"},
)
```

There is no feasible way to directly connect the OAuth2PasswordBearerScopes class to the database for the dynamic lookup of permission sets during the implementation of this project. The only solution is to statically store all these authorization "roles" directly into the constructor of OAuth2PasswordBearerScopes.

Implementing the login transaction

All the scopes will be added to the OAuth2 form login as an option and will be part of the user's login credentials. The following implementation of /ch07/login/token in this new ch07e project shows how to retrieve the scope parameter(s) and the credentials from OAuth2PasswordRequestForm:

```
@router.post("/login/token")
def login(form_data: OAuth2PasswordRequestForm = Depends(),
        sess:Session = Depends(sess_db)):
    username = form_data.username
    password = form_data.password
    loginrepo = LoginRepository(sess)
    account = loginrepo.get_all_login_username(username)
    if authenticate(username, password, account):
        access_token = create_access_token(
            data={"sub": username, "scopes":
              form_data.scopes},
                expires_delta=timedelta(
                minutes=ACCESS_TOKEN_EXPIRE_MINUTES))
        return {"access_token": access_token,
                "token_type": "bearer"}
    else:
        raise HTTPException(
            status_code=400,
            detail="Incorrect username or password")
```

The selected scopes are stored in a list, such as ['user', 'admin_read', 'admin_write', 'bidder_write'], which means that a user has *user*, *administrator (write)*, *administrator (read)*, and *bidder (write)* permissions. create_access_token() will include this list of scopes or "roles" as part of the *payload*, which will be decoded and extracted by get_current_valid_user() through the get_current_user() injectable. By the way, get_current_valid_user() secures every API from the user access by applying the authentication scheme.

Applying the scopes to endpoints

The Security API from the fastapi module replaces the Depends class in injecting the get_current_valid_user() because of its capability to assign scopes to each API service, aside from its capability to perform DI. It has the scopes attribute, where a list of valid scope parameters is defined that restricts the user from access. For instance, the following update_profile() service is accessible only to users whose scopes contain the bidder_write and buyer_write roles:

```
from fastapi.security import SecurityScopes
@router.patch("/profile/update")
def update_profile(id:int, req: ProfileReq,
    current_user: Login = Security(get_current_valid_user,
        scopes=["bidder_write", "buyer_write"]),
    sess:Session = Depends(sess_db)):
    … … … … … …
    if result:
        return JSONResponse(content=
        {'message':'profile updated successfully'},
            status_code=201)
    else:
        return JSONResponse(content=
            {'message':'update profile error'},
                status_code=500)
```

Now, the following code snippet shows the implementation of the get_current_valid_user() injected into every API service by Security:

```
def get_current_valid_user(current_user:
    Login = Security(get_current_user, scopes=["user"])):
    if current_user == None:
        raise HTTPException(status_code=400,
            detail="Invalid user")
    return current_user
```

This method relies on `get_current_user()` when it comes to JWT payload decoding, credential validation, and user scope verification. The user must at least have the `user` scope for the authorization process to proceed. The `Security` class is responsible for injecting `get_current_user()` into `get_current_valid_user()` together with the default `user` scope. Here is the implementation of the `get_current_user()` method:

```
def get_current_user(security_scopes: SecurityScopes,
        token: str = Depends(oauth2_scheme),
            sess:Session = Depends(sess_db)):
    if security_scopes.scopes:
        authenticate_value =
            f'Bearer scope="{security_scopes.scope_str}"'
    else:
        authenticate_value = f"Bearer"
    … … … … … …
    try:
        payload = jwt.decode(token, SECRET_KEY,
                    algorithms=[ALGORITHM])
        username: str = payload.get("sub")
        if username is None:
            raise credentials_exception
        token_scopes = payload.get("scopes", [])
        token_data = TokenData(scopes=token_scopes,
                username=username)
    except JWTError:
        raise credentials_exception
    … … … … … …
    for scope in security_scopes.scopes:
        if scope not in token_data.scopes:
            raise HTTPException(
                status_code=status.HTTP_401_UNAUTHORIZED,
                detail="Not enough permissions",
                headers={"WWW-Authenticate":
                    authenticate_value},
            )
    return user
```

The `SecurityScopes` class of the given `get_current_user()` extracts the scopes assigned to the API service that the user is trying to access. It has a `scope` instance variable that contains all these scope parameters of the API. On the other hand, `token_scopes` carries all the scopes or "roles" of the user extracted from the decoded JWT payload. `get_current_user()` traverses the API scopes in `SecurityScopes` to check whether all of them appear in the `token_scopes` of the user. If `True`, `get_current_user()` authenticates and authorizes the user to access the API service. Otherwise, it throws an change to - exception. The purpose of `TokenData` is to manage the scope parameters from the `token_scopes` payload value and the username.

The next type of OAuth2 authentication scheme that FastAPI can support is the authorization code flow approach.

Building the authorization code flow

If the application is a *public* type and there is no authorization server to process the `client_id` parameter, the `client_secret` parameter, and other related parameters, this OAuth2 authorization code flow approach is appropriate to use. In this scheme, the client creates an authorization request for a short-lived *authorization code* from an `authorizationUrl`. The client will then ask for the token from `tokenUrl` in exchange for the generated code. In this discussion, we will be showcasing another version of our *online auction system* that will use the OAuth2 *authorization code flow* scheme.

Applying OAuth2AuthorizationCodeBearer

The `OAuth2AuthorizationCodeBearer` class is a class from the `fastapi.security` module that builds the authorization code flow. Its constructor requires `authorizationUrl`, `tokenUrl`, and the optional `scopes` before instantiation. The following code shows how this API class is created before its injection into the `get_current_user()` method:

```
from fastapi.security import OAuth2AuthorizationCodeBearer

oauth2_scheme = OAuth2AuthorizationCodeBearer(
    authorizationUrl='ch07/oauth2/authorize',
    tokenUrl="ch07/login/token",
    scopes={"admin_read": "admin ... read only role",
            "admin_write":"admin ... write only role",
            ... ... ... ... ... ...
            "guest":"visitor of the site"},
)
```

The two endpoints, `authorizationUrl` and `tokenUrl`, are crucial parameters in the authentication and authorization process of this scheme. Unlike the previous solutions, we will not rely on the authorization server when generating `access_token`. Instead, we will be implementing an `authorizationUrl` endpoint that will capture essential parameters from the client that will comprise the authorization request for `access_token` generation. The `client_secret` parameter will always remain unexposed to the client.

Implementing the authorization request

In the previous schemes, the `/ch07/login/` token or the `tokenUrl` endpoint is always the redirection point after a login transaction. But this time, the user will be forwarded to the custom `/ch07/oauth2/authorize` or the `authorizationUrl` endpoint for *auth code* generation. Query parameters such as `response_type`, `client_id`, `redirect_uri`, `scope`, and `state` are the essential inputs to the `authorizationUrl` service. The following code from the `/security/secure.py` module of the `ch07f` project will showcase the implementation of the `authorizationUrl` transaction:

```
@router.get("/oauth2/authorize")
def authorizationUrl(state:str, client_id: str,
        redirect_uri: str, scope: str, response_type: str,
        sess:Session = Depends(sess_db)):

    global state_server
    state_server = state

    loginrepo = LoginRepository(sess)
    account = loginrepo.get_all_login_username(client_id)
    auth_code = f"{account.username}:{account.password}
                    :{scope}"
    if authenticate(account.username,
            account.password, account):
        return RedirectResponse(url=redirect_uri
          + "?code=" + auth_code
          + "&grant_type=" + response_type
          + "&redirect_uri=" + redirect_uri
          + "&state=" + state)
    else:
        raise HTTPException(status_code=400,
              detail="Invalid account")
```

These are the query parameters needed by the `authorizationUrl` transaction:

- `response_type`: Custom-generated authorization code
- `client_id`: The public identifier of the app, such as `username`
- `redirect_uri`: The server default URI or a custom endpoint designed to redirect the user back to the application
- `scope`: A scope parameter(s) string, separated by spaces if at least two parameters are involved
- `state`: An arbitrary string value that determines the state of the request

The `redirect_uri` parameter is the destination point where the authentication and authorization processes will occur together with these query parameters.

The generation of `auth_code` is one of the crucial tasks of the `authorizationUrl` transaction, including the authentication process. The *auth code* indicates an ID for the authentication process and is usually unique from all other authentication. There are many ways to generate the code, but in our app, it is simply the combination of user credentials. Conventionally, `auth_code` needs to be encrypted because it comprises the user credentials, scope, and other request-related details.

If the user is valid, the `authorizationUrl` transaction will redirect the user to the `redirect_uri` parameter, back to the FastAPI layer, with the `auth_code`, `grant_type`, and `state` parameters, and the `redirect_uri` parameter itself. The `grant_type` and `redirect_uri` parameters are optional only if the application does not require them. This response will invoke the `tokenUrl` endpoint, which happens to be the `redirectURL` parameter, to pursue the continuation of the authentication process with scoped-based authorization.

Implementing the authorization code response

The `/ch07/login/token` service, or `tokenUrl`, must have the `Form (...)` parameter to capture the `code`, `grant_type`, and `redirect_uri` parameters from the `authorizationUrl` transaction instead of `OAuth2PasswordRequestForm`. The following code snippet shows its implementation:

```
@router.post("/login/token")
def access_token(code: str = Form(...),
  grant_type:str = Form(...), redirect_uri:str = Form(...),
  sess:Session = Depends(sess_db)):
    access_token_expires =
        timedelta(minutes=ACCESS_TOKEN_EXPIRE_MINUTES)

    code_data = code.split(':')
    scopes = code_data[2].split("+")
```

```python
    password = code_data[1]
    username = code_data[0]

    loginrepo = LoginRepository(sess)
    account = loginrepo.get_all_login_username(username)
    if authenticate(username, password, account):
        access_token = create_access_token(
            data={"sub": username, "scopes": scopes},
            expires_delta=access_token_expires,
        )

        global state_server
        state = state_server
        return {
            "access_token": access_token,
            "expires_in": access_token_expires,
            "token_type": "Bearer",
            "userid": username,
            "state": state,
            "scope": "SCOPE"
        }
    else:
        raise HTTPException(
            status_code=400,
            detail="Incorrect credentials")
```

The only response data sent by `authorizationUrl` that is not accessible by `tokenUrl` is the `state` parameter. One workaround is to declare the `state` variable in `authorizationURL` as a `global` one to make it accessible anywhere. The `state` variable is part of the JSON response of the service, which the API authentication requires. Likewise, `tokenUrl` has no access to the user credentials but parsing `auth_code` is a possible way to derive the username, password, and scopes.

If the user is valid, `tokenUrl` must submit the JSON data containing `access_token`, `expires_in`, `token_type`, `userid`, and `state` to proceed with the authentication scheme.

This authorization code flow scheme provides the baseline protocol for the *OpenID Connect* authentication. Various identity and access management solutions, such as *Okta*, *Auth0*, and *Keycloak*, apply the authorization requests and responses involving response_type code. The next topic will highlight the FastAPI's support of the OpenID Connect specification.

Applying the OpenID Connect specification

There are three *online auction* projects created to impose the *OAuth2 OpenID Connect* authentication scheme. All these projects use third-party tools to perform authentication and authorization procedures. The ch07g project uses *Auth0*, ch07h uses *Okta*, and ch07i applies a *Keycloak* policy in authenticating client access to the API services. Let us first highlight Keycloak's support for the OpenID Connect protocol.

Using HTTPBearer

The HTTPBearer class is a utility class from the fastapi.security module that provides an authorization scheme that relies directly on the authorization header with the Bearer tokens. Unlike the other OAuth2 schemes, this requires the generation of an access_token on the *Keycloak* side before running the authentication server. At this point, the framework has no straightforward way of accessing the credentials and the access_token from Keycloak's identity provider. To utilize this class, we only need to instantiate it without any constructor parameters.

Installing and configuring the Keycloak environment

Keycloak is a Java-based application that we can download from the following link: https://www.keycloak.org/downloads. After downloading, we can unzip its content to any directory. But before running it, we need to install at least the Java 12 SDK on our development machine. Once you have completed the setup, run its bin\standalone.bat or bin\standalone.sh on the console and then open http://localhost:8080 on the browser. Afterward, create an administration account to set up the *realm*, *clients*, *users*, and *scopes*.

Setting the Keycloak realm and clients

A Keycloak *realm* is an object that encompasses all the clients together with their *credentials*, *scopes*, and *roles*. The first step before creating the user profiles is to build a realm, as shown in *Figure 7.7*:

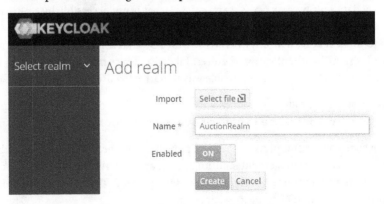

Figure 7.7 - Creating a Keycloak realm

After the realm, the Keycloak *client*, which manages the user profiles and credentials, is the next priority. It is created on the **Configure | Clients** panel, as shown:

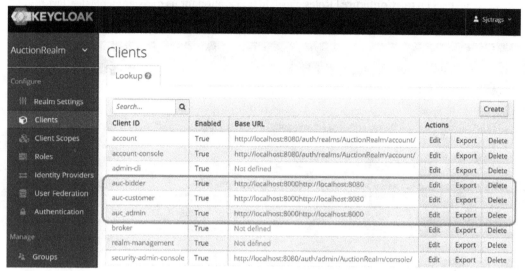

Figure 7.8 – Creating the Keycloak clients

After creating the clients, we need to edit each Client profile to input the following details:

- Its access type must be `confidential`

- `Authorization Enabled` is turned `ON`

- Provide values for `Root URL`, `Base URL`, and `Admin URL`, which all refer to the `http://localhost:8000` of the API service application

- Specify a `Valid Redirect URI` endpoint, or we can just assign `http://localhost:8080/*` if we have no specific custom endpoint

- In `Advanced Settings`, set `Access Token Lifespan` (e.g., 15 minutes)

- Under `Authentication Flow Overrides`, set `Browser Flow` to `browser` and `Direct Grant Flow` to `direct grant`.

In the **Credentials** panel, we can find the *client credentials*, in which the auto-generated `client_secret` value is located. After the setup, we can now assign users to the clients.

Creating users and user roles

First, we create *roles* on the **Configure | Roles** panel, in preparation for user assignment later. *Figure 7.9* shows three user roles that will handle the application's *administration*, *auctioning*, and *bidding* tasks:

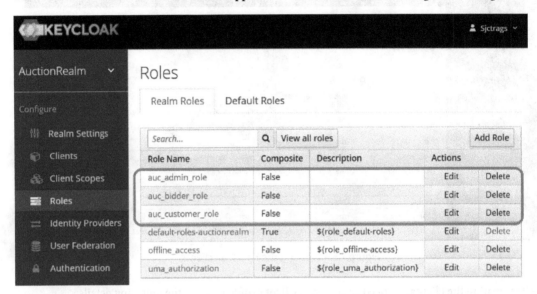

Figure 7.9 – Creating user roles

After creating the roles, we need to build the list of users on the **Manage | Users** panel. *Figure 7.10* shows the three created users, each with the mapped roles:

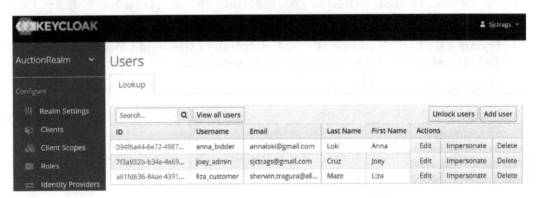

Figure 7.10 – Creating client users

To provide users with their roles, we need to click the **Edit** link for each user and assign the appropriate **Realm Roles**. *Figure 7.11* shows that the user `joey_admin` has the `auc_admin_role` role, authorizing the user to do the administrative tasks for the app. By the way, do not forget to create a password for each user on the **Credentials** panel:

Figure 7.11 – Mapping user roles

Assigning roles to clients

Aside from user roles, clients can also have assigned roles. A *client role* defines the type of users a client must have under its coverage. It also provides the client's boundary when accessing the API services. *Figure 7.12* shows `auc_admin` with an `admin` role:

Figure 7.12 – Creating client roles

Then, we need to return to the **Manage | Users** panel and assign the *user* its role(s) through the *client*. For instance, *Figure 7.13* shows that joey_admin has the admin role because the auc_admin role was added to its profile. All users with the auc_admin client added to their setup have *admin* access to the app, including joey_admin:

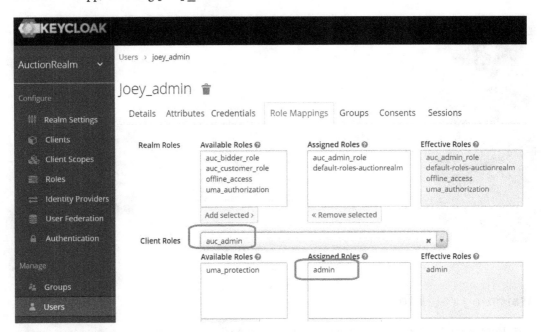

Figure 7.13 – Mapping client roles to users

Creating user permissions through scopes

To assign permission to each client, we need to create *client scopes* on the **Configure | Client Scopes** panel. Each client scope must have an Audience-type token mapper. *Figure 7.14* shows the admin:read and admin:write scopes for the auc_admin client, auction:read and auction:write for auc_customer, and bidder:write and bidder:read for auc_bidder:

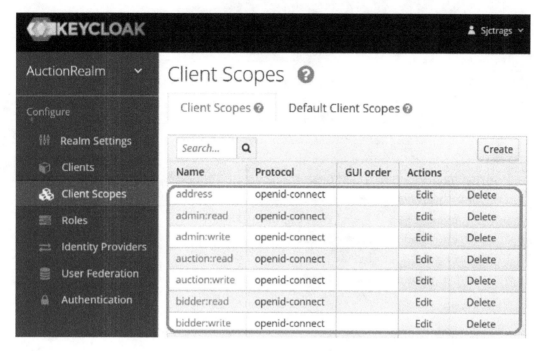

Figure 7.14 – Creating the client scopes

These *client scopes* are essential details within the `Security` injection for each API service if *scope-based authorization* is part of the scheme.

Integrating Keycloak with FastAPI

Since the FastAPI application cannot directly access the Keycloak client credentials for authentication, the application has a `login_keycloak()` service to redirect the user to the *AuctionRealm URI*, our custom `authorizationUrl` in Keycloak. The URI is `/auth/realms/AuctionRealm/protocol/openid-connect/auth`. First, access `http://localhost:8080/auth/realms/AuctionRealm/account/` to log in using the authorized user credentials, such as `joey_admin`, before invoking the `login_keycloak()` service.

Now, the redirection must include `client_id`, as with the `auc_admin` client, and the custom callback handler called `redirect_uri`. All the Keycloak realm details must be in the `.config` property file. The following code shows the implementation of the `login_keycloak()` service:

```
import hashlib
import os
import urllib.parse as parse
```

```
@router.get("/auth/login")
def login_keycloak() -> RedirectResponse:
    config = set_up()
    state = hashlib.sha256(os.urandom(32)).hexdigest()

    AUTH_BASE_URL = f"{config['KEYCLOAK_BASE_URL']}
     /auth/realms/AuctionRealm/protocol/
        openid-connect/auth"
    AUTH_URL = AUTH_BASE_URL +
     '?{}'.format(parse.urlencode({
        'client_id': config["CLIENT_ID"],
        'redirect_uri': config["REDIRECT_URI"],
        'state': state,
        'response_type': 'code'
    }))

    response = RedirectResponse(AUTH_URL)
    response.set_cookie(key="AUTH_STATE", value=state)
    return response
```

A *state* is part of `login_keycloak()`'s response for the callback method to verify the authentication, a similar approach we had in utilizing `OAuth2AuthorizationCodeBearer`. The service used the `hashlib` module to generate a randomized hash string value for the state using the *SHA256* encryption algorithm. On the other hand, Keycloak's *AuctionRealm URI* must return a JSON result as follows:

```
{"access_token":"eyJhbGciOiJSUzI1NiIsInR5cCIgOiAiSldUIiwia2lkI
iA6ICJJMFR3YVhiZnh0MVNQSnNzVTByQ09hMzVDaTdZNDkzUnJIeDJTM3paa0V
VIn0.eyJleHAiOjE2NDU0MTgzNTAsImlhdCI6MTY0NTQxNzQ1MCwiYXV0aF90a
W1lIjoxNjQ1NDE3NDM3LCJqdGkiOiI4YTQzMjBmBmYi0xMzg5LTQ2NzU..........................
.........2YTU2In0.UktwOX7H2ZdoyP1VZ5V2MXUX2Gj41D2cuusvwEZXBtVMvnoTDh
KJgN8XWL7P3ozv4A1ZlBmy4NX1HHjPbSGsp2cvkAWwlyXmhyUzfQslf8Su00-4
e9FR4i4rOQtNQfqHM7cLhrzr3-od-uyj1m9KsrpbqdLvPEl3KZnmOfFbTwUXfE
9YclBFa8zwytEWb4qvLvKrA6nPv7maF2_MagMD_0Mh9t95N9_aY9dfquS9tcEV
Whr3d9B3ZxyOtjO8WiQSJyjLCT7IW1hesa8RL3WsiG3QQQ4nUKVHhnciK8efRm
XeaY6iZ_-8jm-mqMBxw00-jchJE8hMtLUPQTMIK0eopA","expires_in":900,
"refresh_expires_in":1800,"refresh_token":"eyJhbGciOiJIUzI1NiIs
InR5cCIgOiAiSldUIiwia2lkIiA6ICJhNmVmZGQ0OS0yZDIxLTQ0NjQtOGUyOC0
4ZWJkMjdiZjFmOTkifQ.eyJleHAiOjE2NDU0MTkyNTAsImlhdCI6MTY0NTQxNzQ
1MCwianRpIjoiMzRiZmMzMmYtYjAzYi00MDM3LTk5YzMt.........................zc2lvbl9z
```

dGF0ZSI6ImM1NTE3ZDIwLTMzMTgtNDFlMi1hNTlkLWU2MGRiOWM1NmE1NiIsIn
Njb3BlIjoiYWRtaW46d3JpdGUgYWRtaW46cmVhZCB1c2VyIiwic2lkIjoiYzU1
MTdkMjAtMzMxOC00MWUyLWE1OWQtZTYwZGI5YzU2YTU2In0.xYYQPr8dm7_o1G
KplnS5cWmLbpJTCBDfm1WwZLBhM6k","token_type":"Bearer","not-
before-policy":0,"session_state":"c5517d20-3318-41e2-a59d-e60d
b9c56a56","scope":"admin:write admin:read user"}

This contains the essential credentials, such as `access_token`, `expires_in`, `session_state`, and `scope`.

Implementing the token verification

The application's `HTTPBearer` needs `access_token` to pursue the client-side authentication. On the OpenAPI dashboard, we click the **Authorize** button and paste the `access_token` value provided by Keycloak's `authorizationUrl`. After the successful authentication, `get_current_user()` will verify the access to each API endpoint based on the credentials extracted from `access_token`. The following code highlights the `get_current_user()`, which builds the user credentials from Keycloak's *token* using the `PyJWT` utility and algorithms such as `RSAAlgorithm`:

```python
from jwt.algorithms import RSAAlgorithm
from urllib.request import urlopen
import jwt

def get_current_user(security_scopes: SecurityScopes,
        token: str = Depends(token_auth_scheme)):
    token = token.credentials
    config = set_up()
    jsonurl = urlopen(f'{config["KEYCLOAK_BASE_URL"]}
        /auth/realms/AuctionRealm/protocol
        /openid-connect/certs')
    jwks = json.loads(jsonurl.read())
    unverified_header = jwt.get_unverified_header(token)

    rsa_key = {}
    for key in jwks["keys"]:
        if key["kid"] == unverified_header["kid"]:
            rsa_key = {
```

```
            "kty": key["kty"],
            "kid": key["kid"],
            "use": key["use"],
            "n": key["n"],
            "e": key["e"]
        }

if rsa_key:
    try:
            public_key = RSAAlgorithm.from_jwk(rsa_key)
            payload = jwt.decode(
                token,
                public_key,
                algorithms=config["ALGORITHMS"],
                options=dict(
                        verify_aud=False,
                        verify_sub=False,
                        verify_exp=False,
                    )
                )
… … … … … …
token_scopes = payload.get("scope", "").split()

for scope in security_scopes.scopes:
    if scope not in token_scopes:
        raise AuthError(
            {
            "code": "Unauthorized",
            "description": Invalid Keycloak details,
            },403,
        )
    return payload
```

Install the PyJWT module first to utilize the needed encoders and decoder functions. The jwt module has RSAAlgorithm, which can help decode the rsa_key from the token with some options disabled, such as the verification of the client's audience.

Integrating Auth0 with FastAPI

Auth0 can also be a third-party authentication provider that can authenticate and authorize access to the API endpoints of our app. But first, we need to sign up for an account at `https://auth0.com/`.

After signing up for an account, create an Auth0 application to derive **Domain**, **Client ID**, and **Client Secret**, and configure some URI- and token-related details. *Figure 7.15* shows the dashboard that creates the Auth0 application:

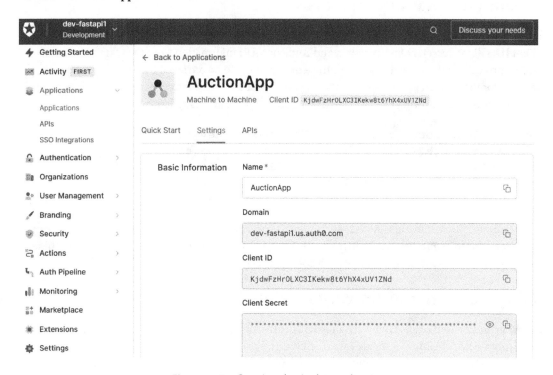

Figure 7.15 – Creating the Auth0 application

The Auth0 application also has the generated *Audience API* URI that the client-side authentication needs. On the other hand, part of the authentication parameters is the *issuer*, which we can derive from the **Domain** value of the Auth0 application. The issuer is a base URI to the `/oauth/token` service that generates the `auth_token` once requested, similar to the Keycloak's realm. We place all these Auth0 details in the `.config` file, including the PyJWT algorithm for decoding `auth_token`.

`ch07g` has its own version of `get_current_user()` that processes the *payload* for API authentication and authorization from the Auth0 details in the `.config` file. But first, the HTTPBearer class needs the `auth_token` value and gets it by running the following `tokenURL` of our Auth0 application, *AuctionApp*:

```
curl --request POST                                        --url
https://dev-fastapi1.us.auth0.com/oauth/token    --header
```

```
'content-type: application/json'                    --data "{"client_
id":"KjdwFzHrOLXC3IKe
kw8t6YhX4xUV1ZNd",   "client_secret":"_
KyPEUOB7DA5Z3mmRXpnqWA3EXfrjLw2R5SoUW7m1wLMj7
KoElMyDLiZU8SgMQYr","audience":"https://fastapi.auction.com/",
"grant_type":"client_credentials"}"
```

Integrating Okta with FastAPI

Some of the processes performed in Auth0 are also found in Okta's procedures when extracting the **Client ID**, **Client Secret**, **Domain**, issuer, and Audience API information from the Okta account. The ch07h project has these details stored in the app.env file to be retrieved by its get_current_user() for the payload generation. But then again, the HTTPBearer class needs an auth_token from executing the following Okta's tokenURL, based on the account's issuer:

```
curl --location --request POST "https://dev-5180227.
okta.com/oauth2/default/v1/token?grant_type=client_
credentials&client_id=0oa3tvejee5UPt7QZ5d7&client_
secret=LA4WP8lACWKu4Ke9fReol0fNSUvxsxTvGLZdDS5-"    --header
"Content-Type: application/x-www-form-urlencoded"
```

Aside from the Basic, Digest, OAuth2, and OpenID Connect authentication schemes, FastAPI can utilize some built-in middlewares to help secure API endpoints. Let us now determine whether these middlewares can provide a custom authentication process.

Using built-in middlewares for authentication

FastAPI can use Starlette middleware such as AuthenticationMiddleware to implement any custom authentication. It needs AuthenticationBackend to implement the scheme for our app's security model. The following custom AuthenticationBackend checks whether the Authorization credential is a Bearer class and verifies whether the username token is equivalent to a fixed username credential provided by the middleware:

```
class UsernameAuthBackend(AuthenticationBackend):
    def __init__(self, username):
        self.username = username

    async def authenticate(self, request):
        if "Authorization" not in request.headers:
            return
        auth = request.headers["Authorization"]
        try:
```

```
        scheme, username = auth.split()
        if scheme.lower().strip() != 'bearer'.strip():
            return
    except:
        raise AuthenticationError(
            'Invalid basic auth credentials')
    if not username == self.username:
        return

    return AuthCredentials(["authenticated"]),
        SimpleUser(username)
```

Activating this `UsernameAuthBackend` means injecting it into the FastAPI constructor in `main.py` with `AuthenticationMiddleware`. It also needs the designated `username` for its authentication process to work. The following snippet shows how to activate the whole authentication scheme in the `main.py` file:

```
from security.secure import UsernameAuthBackend
from starlette.middleware import Middleware
from starlette.middleware.authentication import
    AuthenticationMiddleware

middleware = [Middleware(AuthenticationMiddleware,
    backend=UsernameAuthBackend("sjctrags"))]
app = FastAPI(middleware=middleware)
```

Injecting FastAPI's `Request` is the first step in applying the authentication scheme. Then, we decorate each API with `@requires("authenticated")` after the `@router` decorator. We can extend the `UsernameAuthBackend` process further by adding JWT encoding and decoding, encryption, or custom roles-based authorization.

Summary

Securing any applications is always the main priority in producing quality software. We always choose frameworks that support reliable and credible security solutions, and that can at least prevent malicious attacks from the outside environment. Although we know for a fact that a perfect security model is a myth, we always develop security solutions that can cope with the threats we know.

FastAPI is one of the API frameworks that has built-in support for many popular authentication processes, from Basic to the OpenID Connect specification. It fully supports all effective OAuth2 authentication schemes and is even open to further customization of its security APIs.

Although it has no direct support for the OpenID Connect specification, it can still integrate seamlessly with different popular identities and user management systems, such as Auth0, Okta, and Keycloak. This framework may still surprise us with many security utilities and classes in the future that we can apply to build scalable microservice applications.

The next chapter will focus on topics regarding non-blocking API services, events, and message-driven transactions.

8

Creating Coroutines, Events, and Message-Driven Transactions

The FastAPI framework is an asynchronous framework that runs over the asyncio platform, which utilizes the ASGI protocol. It is well known for its 100% support for asynchronous endpoints and non-blocking tasks. This chapter will focus on how we create highly scalable applications with asynchronous tasks and event-driven and message-driven transactions.

We learned in *Chapter 2, Exploring the Core Features*, that *Async/Await* or asynchronous programming is a design pattern that enables other services or transactions to run outside the main thread. The framework uses the `async` keyword to create asynchronous processes that will run on top of other thread pools and will be *awaited*, instead of invoking them directly. The number of external threads is defined during the Uvicorn server startup through the `--worker` option.

In this chapter, we will delve into the framework and scrutinize the various components of the FastAPI Framework that can run asynchronously using multiple threads. The following highlights will help us understand how asynchronous FastAPI is:

- Implementing coroutines
- Creating asynchronous background tasks
- Understanding Celery tasks
- Building message-driven transactions using RabbitMQ
- Building publish/subscribe messaging using Kafka
- Applying reactive programming in tasks
- Customizing events

- Implementing asynchronous **Server-Sent Events (SSE)**
- Building an asynchronous WebSocket

Technical requirements

This chapter will cover asynchronous features, software specifications, and the components of a *newsstand management system* prototype. The discussions will use this online newspaper management system prototype as a specimen to understand, explore, and implement asynchronous transactions that will manage the *newspaper content, subscription, billing, user profiles, customers*, and other business-related transactions. The code has all been uploaded to `https://github.com/PacktPublishing/Building-Python-Microservices-with-FastAPI` under the ch08 project.

Implementing coroutines

In the FastAPI framework, a *thread pool* is always present to execute both synchronous API and non-API transactions for every request. For ideal cases where both the transactions have minimal performance overhead with *CPU-bound* and *I/O-bound* transactions, the overall performance of using the FastAPI framework is still better than those frameworks that use non-ASGI-based platforms. However, when contention occurs due to high CPU-bound traffic or heavy CPU workloads, the performance of FastAPI starts to wane due to *thread switching*.

Thread switching is a context switch from one thread to another within the same process. So, if we have several transactions with varying workloads running in the background and on the browser, FastAPI will run these transactions in the thread pool with several context switches. This scenario will cause contention and degradation to lighter workloads. To avoid performance issues, we apply *coroutine switching* instead of threads.

Applying coroutine switching

The FastAPI framework works at the optimum speed with a mechanism called *coroutine switching*. This approach allows transaction-tuned tasks to work cooperatively by allowing other running processes to pause so that the thread can execute and finish more urgent tasks, and resume "awaited" transactions without preempting the thread. These coroutine switches are programmer-defined components and not kernel-related or memory-related features. In FastAPI, there are two ways of implementing coroutines: (a) applying the `@asyncio.coroutine` decorator, and (b) using the `async/await` construct.

Applying @asyncio.coroutine

`asyncio` is a Python extension that implements the Python concurrency paradigm using a single-threaded and single-process model and provides API classes and methods for running and managing coroutines. This extension provides an `@asyncio.coroutine` decorator that transforms API and native services into *generator-based coroutines*. However, this is an old approach and can only be used in FastAPI that uses Python 3.9 and below. The following is a login service transaction of our *newsstand management system* prototype implemented as a coroutine:

```python
@asyncio.coroutine
def build_user_list(query_list):
    user_list = []
    for record in query_list:
        yield from asyncio.sleep(2)
        user_list.append(" ".join([str(record.id),
            record.username, record.password]))
    return user_list
```

`build_user_list()` is a native service that converts all login records into the `str` format. It is decorated with the `@asyncio.coroutine` decorator to transform the transaction into an asynchronous task or coroutine. A coroutine can invoke another coroutine function or method using only the `yield from` clause. This construct pauses the coroutine and passes the control of the thread to the coroutine function invoked. By the way, the `asyncio.sleep()` method is one of the most widely used asynchronous utilities of the `asyncio` module, which can pause a process for a few seconds, but is not the ideal one. On the other hand, the following code is an API service implemented as a coroutine that can minimize contention and performance degradation in client-side executions:

```python
@router.get("/login/list/all")
@asyncio.coroutine
def list_login():
    repo = LoginRepository()
    result = yield from repo.get_all_login()
    data = jsonable_encoder(result)
    return data
```

The `list_login()` API service retrieves all the login details of the application's users through a coroutine CRUD transaction implemented in *GINO ORM*. The API service again uses the `yield from` clause to run and execute the `get_all_login()` coroutine function.

A coroutine function can invoke and await multiple coroutines concurrently using the `asyncio.gather()` utility. This `asyncio` method manages a list of coroutines and waits until all its coroutines have completed their tasks. Then, it will return a list of results from the corresponding coroutines. The following code is an API that retrieves login records through an asynchronous CRUD transaction and then invokes `count_login()` and `build_user_list()` concurrently to process these records:

```
@router.get("/login/list/records")
@asyncio.coroutine
def list_login_records():
    repo = LoginRepository()
    login_data = yield from repo.get_all_login()
    result = yield from
        asyncio.gather(count_login(login_data),
            build_user_list(login_data))
    data = jsonable_encoder(result[1])
    return {'num_rec': result[0], 'user_list': data}
```

`list_login_records()` uses `asyncio.gather()` to run the `count_login()` and `build_user_list()` tasks and later extract their corresponding returned values for processing.

Using the async/await construct

Another way of implementing a coroutine is using `async/await` constructs. As with the previous approach, this syntax creates a task that can pause anytime during its operation before it reaches the end. But the kind of coroutine that this approach produces is called a *native coroutine*, which is not iterable in the way that the generator type is. The `async/await` syntax also allows the creation of other asynchronous components such as the `async with` context managers and `async for` iterators. The following code is the `count_login()` task previously invoked in the generator-based coroutine service, `list_login_records()`:

```
async def count_login(query_list):
    await asyncio.sleep(2)
    return len(query_list)
```

The `count_login()` native service is a native coroutine because of the `async` keyword placed before its method definition. It only uses `await` to invoke other coroutines. The `await` keyword suspends the execution of the current coroutine and passes the control of the thread to the invoked coroutine function. After the invoked coroutine finishes its process, the thread control will yield back to the caller coroutine. Using the `yield from` construct instead of `await` will raise an error because our coroutine here is not generator-based. The following is an API service implemented as a native coroutine that manages data entry for the new administrator profiles:

```
@router.post("/admin/add")
async def add_admin(req: AdminReq):
    admin_dict = req.dict(exclude_unset=True)
    repo = AdminRepository()
    result = await repo.insert_admin(admin_dict)
    if result == True:
        return req
    else:
        return JSONResponse(content={'message':'update
            trainer profile problem encountered'},
            status_code=500)
```

Both generator-based and native coroutines are monitored and managed by an *event loop*, which represents an infinite loop inside a thread. Technically, it is an object found in the *thread*, and each *thread* in the thread pool can only have one event loop, which contains a list of helper objects called *tasks*. Each task, pre-generated or manually created, executes one coroutine. For instance, when the previous `add_admin()` API service invokes the `insert_admin()` coroutine transaction, the event loop will suspend `add_admin()` and tag its task as an *awaited* task. Afterward, the event loop will assign a task to run the `insert_admin()` transaction. Once the task has completed its execution, it will yield the control back to `add_admin()`. The thread that manages the FastAPI application is not interrupted during these shifts of execution since it is the event loop and its tasks that participate in the *coroutine switching* mechanism. Let us now use these coroutines to build our application

Designing asynchronous transactions

There are a few programming paradigms that we can follow when creating coroutines for our application. Utilizing more coroutine switching in the process can help improve the software performance. In our *newsstand* application, there is an endpoint, /admin/login/list/enc, in the admin.py router that returns a list of encrypted user details. In its API service, shown in the following code, each record is managed by an extract_enc_admin_profile() transaction call instead of passing the whole data record to a single call, thus allowing the concurrent executions of tasks. This strategy is better than running the bulk of transactions in a thread without *context switches*:

```python
@router.get("/admin/login/list/enc")
async def generate_encypted_profile():
    repo = AdminLoginRepository()
    result = await repo.join_login_admin()
    encoded_data = await asyncio.gather(
        *(extract_enc_admin_profile(rec) for rec in result))
    return encoded_data
```

Now, the extract_enc_admin_profile() coroutine, shown in the following code, implements a chaining design pattern, where it calls the other smaller coroutines through a chain. Simplifying and breaking down the monolithic and complex processes into smaller but more robust coroutines will improve the application's performance by utilizing more context switches. In this API, extract_enc_admin_profile() creates three context switches in a chain, better than thread switches:

```python
async def extract_enc_admin_profile(admin_rec):
    p = await extract_profile(admin_rec)
    pinfo = await extract_condensed(p)
    encp = await decrypt_profile(pinfo)
    return encp
```

On the other hand, the following implementation is the smaller subroutines awaited and executed by extract_enc_admin_profile():

```python
async def extract_profile(admin_details):
    profile = {}
    login = admin_details.parent
    profile['firstname'] = admin_details.firstname
    ... ... ... ... ... ...
    profile['password'] = login.password
    await asyncio.sleep(1)
```

```
        return profile

async def extract_condensed(profiles):
    profile_info = " ".join([profiles['firstname'],
        profiles['lastname'], profiles['username'],
        profiles['password']])
    await asyncio.sleep(1)
    return profile_info

async def decrypt_profile(profile_info):
    key = Fernet.generate_key()
    fernet = Fernet(key)
    encoded_profile = fernet.encrypt(profile_info.encode())
    return encoded_profile
```

These three subroutines will give the main coroutine the encrypted `str` that contains the details of an administrator profile. All these encrypted strings will be collated by the API service using the `asyncio.gather()` utility.

Another programming approach to utilizing the coroutine switching is the use of pipelines created by `asyncio.Queue`. In this programming design, the queue structure is the common point between two tasks: (a) the task that will place a value to the queue called the *producer*, and (b) the task that will fetch the item from the queue, the *consumer*. We can implement a *one producer/one consumer* interaction or a *multiple producers/multiple consumers* setup with this approach.

The following code highlights the `process_billing()` native service that builds a *producer/consumer* transaction flow. The `extract_billing()` coroutine is the producer that retrieves the billing records from the database and passes each record one at a time to the queue. `build_billing_sheet()`, on the other hand, is the consumer that fetches the record from the queue structure and generates the billing sheet:

```
async def process_billing(query_list):
    billing_list = []

    async def extract_billing(qlist, q: Queue):
        assigned_billing = {}
        for record in qlist:
            await asyncio.sleep(2)
            assigned_billing['admin_name'] = "{} {}"
                .format(record.firstname, record.lastname)
```

```python
        if not len(record.children) == 0:
            assigned_billing['billing_items'] =
                record.children
        else:
            assigned_billing['billing_items'] = None


    await q.put(assigned_billing)

async def build_billing_sheet(q: Queue):
    while True:
        await asyncio.sleep(2)
        assigned_billing = await q.get()
        name = assigned_billing['admin_name']
        billing_items =
            assigned_billing['billing_items']
        if not billing_items == None:
            for item in billing_items:
                billing_list.append(
                    {'admin_name': name, 'billing': item})
        else:
            billing_list.append(
                    {'admin_name': name, 'billing': None})
        q.task_done()
```

In this programming design, the build_billing() coroutine will explicitly wait for the record queued by extract_billing(). This setup is possible due to the asyncio.create_task() utility, which directly assigns and schedules a task to each coroutine.

The queue is the only method parameter common to the coroutines because it is their common point. The join() of asyncio.Queue ensures that all the items passed to the pipeline by extract_billing() are fetched and processed by build_billing_sheet(). It also blocks the external controls that would affect the coroutine interactions. The following code shows how to create asyncio.Queue and schedule a task for execution:

```python
q = asyncio.Queue()
build_sheet = asyncio.create_task(
        build_billing_sheet(q))
await asyncio.gather(asyncio.create_task(
        extract_billing(query_list, q)))
```

```
    await q.join()
    build_sheet.cancel()
    return billing_list
```

By the way, always pass `cancel()` to the task right after its coroutine has completed the process. On the other hand, we can also apply other ways so that the performance of our coroutines can improve.

Using the HTTP/2 protocol

Coroutine execution can be faster in applications running on the *HTTP/2* protocol. We can replace the *Uvicorn* server with *Hypercorn*, which now supports ASGI-based frameworks such as FastAPI. But first, we need to install `hypercorn` using `pip`:

```
pip install hypercorn
```

For *HTTP/2* to work, we need to create an SSL certificate. Using OpenSSL, our app has two *PEM* files for our *newsstand* prototype: (a) the private encryption (`key.pem`) and (b) the certificate information (`cert.pem`.) We place these files in the main project folder before executing the following `hypercorn` command to run our FastAPI application:

```
hypercorn --keyfile key.pem --certfile cert.pem main:app
--bind 'localhost:8000' --reload
```

Now, let us explore other FastAPI tasks that can also use coroutines.

Creating asynchronous background tasks

In *Chapter 2, Exploring the Core Features*, we first showcased the `BackgroundTasks` injectable API class, but we didn't mention creating asynchronous background tasks. In this discussion, we will be focusing on creating asynchronous background tasks using the `asyncio` module and coroutines.

Using the coroutines

The framework supports the creation and execution of asynchronous background processes using the `async/await` structure. The following native service is an asynchronous transaction that generates a billing sheet in CSV format in the background:

```
async def generate_billing_sheet(billing_date, query_list):
    filepath = os.getcwd() + '/data/billing-' +
                str(billing_date) +'.csv'
    with open(filepath, mode="a") as sheet:
        for vendor in query_list:
```

```
            billing = vendor.children
            for record in billing:
                if billing_date == record.date_billed:
                    entry = ";".join(
                [str(record.date_billed), vendor.account_name,
                 vendor.account_number, str(record.payable),
                 str(record.total_issues) ])
                    sheet.write(entry)
            await asyncio.sleep(1)
```

This `generate_billing_sheet()` coroutine service will be executed as a background task in the following API service, `save_vendor_billing()`:

```
@router.post("/billing/save/csv")
async def save_vendor_billing(billing_date:date,
                tasks: BackgroundTasks):
    repo = BillingVendorRepository()
    result = await repo.join_vendor_billing()
    tasks.add_task(generate_billing_sheet,
            billing_date, result)
    tasks.add_task(create_total_payables_year,
            billing_date, result)
    return {"message" : "done"}
```

Now, nothing has changed when it comes to defining background processes. We usually inject `BackgroundTasks` into the API service method and apply `add_task()` to provide task schedules, assignments, and execution for a specific process. But since the approach is now to utilize coroutines, the background task will use the event loop instead of waiting for the current thread to finish its jobs.

If the background process requires arguments, we pass these arguments to `add_task()` right after its *first parameter*. For instance, the arguments for the `billing_date` and `query_list` parameters of `generate_billing_sheet()` should be placed after the `generate_billing_sheet` injection into `add_task()`. Moreover, the `billing_date` value should be passed before the `result` argument because `add_task()` still follows the order of parameter declaration in `generate_billing_sheet()` to avoid a type mismatch.

All asynchronous background tasks will continuously execute and will not be *awaited* even if their coroutine API service has already returned a response to the client.

Creating multiple tasks

`BackgroundTasks` allows the creation of multiple asynchronous transactions that will execute concurrently in the background. In the `save_vendor_billing()` service, there is another task created for a new transaction called the `create_total_payables_year()` transaction, which requires the same arguments as `generate_billing_sheet()`. Again, this newly created task will be utilizing the event loop instead of the thread.

The application always encounters performance issues when the background processes have high-CPU workloads. Also, tasks generated by `BackgroundTasks` are not capable of returning values from the transactions. Let us look for another solution where tasks can manage high workloads and execute processes with returned values.

Understanding Celery tasks

Celery is a non-blocking task queue that runs on a distributed system. It can manage asynchronous background processes that are huge and heavy with CPU workloads. It is a third-party tool, so we need to install it first through `pip`:

```
pip install celery
```

It schedules and runs tasks concurrently on a single server or distributed environment. But it requires a message transport to send and receive messages, such as *Redis*, an in-memory database that can be used as a message broker for messages in strings, dictionaries, lists, sets, bitmaps, and stream types. Also, we can install Redis on Linux, macOS, and Windows. Now, after the installation, run its `redis-server.exe` command to start the server. In Windows, the Redis service is set to run by default after installation, which causes a *TCP bind listener* error. So, we need to stop it before running the startup command. *Figure 8.1* shows Windows **Task Manager** with the Redis service giving a **Stopped** status:

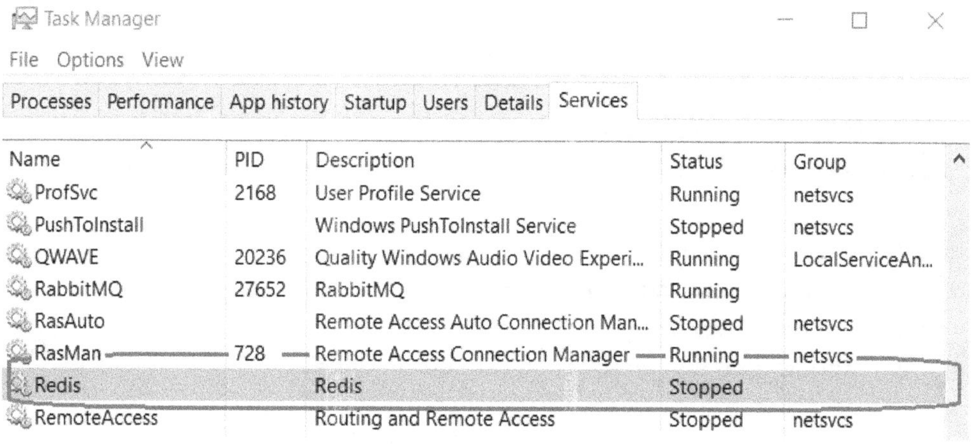

Figure 8.1 – Stopping the Redis service

After stopping the service, we should now see Redis running as shown in *Figure 8.2*:

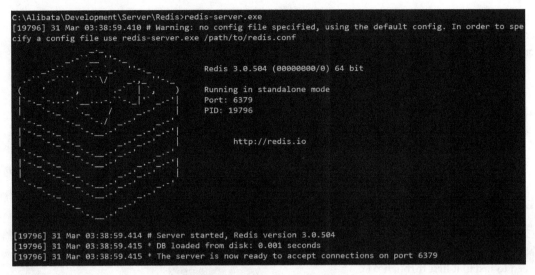

Figure 8.2 – A running Redis server

Creating and configuring the Celery instance

Before creating Celery tasks, we need a Celery instance placed in a dedicated module of our application. The *newsstand* prototype has the Celery instance in the `/services/billing.py` module, and the following is part of the code that shows the process of Celery instantiation:

```python
from celery import Celery
from celery.utils.log import import get_task_logger

celery = Celery("services.billing",
    broker='redis://localhost:6379/0',
    backend='redis://localhost',
    include=["services.billing", "models", "config"])

class CeleryConfig:
    task_create_missing_queues = True
    celery_store_errors_even_if_ignored = True
    task_store_errors_even_if_ignored = True
```

```
        task_ignore_result = False
        task_serializer = "pickle"
        result_serializer = "pickle"
        event_serializer = "json"
        accept_content = ["pickle", "application/json",
                "application/x-python-serialize"]
        result_accept_content = ["pickle", "application/json",
                "application/x-python-serialize"]
  celery.config_from_object(CeleryConfig)


  celery_log = get_task_logger(__name__)
```

To create the Celery instance, we need the following details:

- The name of the current module containing the Celery instance (the first argument)

- The URL of Redis as our message broker (`broker`)

- The backend result where the results of tasks are stored and monitored (`backend`)

- The list of other modules used in the message body or by the Celery task (`include`)

After the instantiation, we need to set the appropriate serializer and content types to process the incoming and outgoing message body of the tasks involved, if there are any. To allow the passing of full Python objects with non-JSON-able values, we need to include `pickle` as a supported content type, then declare a default task and result serializer to the object stream. However, using a `pickle` serializer poses some security issues because it tends to expose some transaction data. To avoid compromising the app, apply sanitation to message objects, such as removing sensitive values or credentials, before pursuing the messaging operation.

Apart from the serialization options, other important properties such as `task_create_missing_queues`, `task_ignore_result`, and error-related configuration should also be part of the `CeleryConfig` class. Now, we declare all these details in a custom class, which we will inject into the `config_from_object()` method of the Celery instance.

Additionally, we can create a Celery logger through its `get_task_logger()` with the name of the current task.

Creating the task

The main goal of the *Celery instance* is to annotate Python methods to become tasks. The Celery instance has a `task()` decorator that we can apply to all callable procedures we want to define as asynchronous tasks. Part of the `task()` decorator is the task's `name`, an optional unique name composed of the *package, module name(s),* and the *method name of the transaction.* It has other attributes that can add more refinement to the task definition, such as the `auto_retry` list, which registers `Exception` classes that may cause execution retries when emitted, and `max_tries`, which limits the number of retry executions of a task. By the way, Celery 5.2.3 and below can only define tasks from *non-coroutine methods.*

The `services.billing.tasks.create_total_payables_year_celery` task shown here adds all the payable amounts per date and returns the total amount:

```python
@celery.task(
    name="services.billing.tasks
            .create_total_payables_year_celery",
                auto_retry=[ValueError, TypeError],
                max_tries=5)
def create_total_payables_year_celery(billing_date,
                query_list):
        total = 0.0
        for vendor in query_list:
            billing = vendor.children
            for record in billing:
                if billing_date == record.date_billed:
                    total += record.payable
        celery_log.info('computed result: ' + str(total))
        return total
```

The given task has only five (5) retries to recover when it encounters either `ValueError` or `TypeError` at runtime. Also, it is a function that returns a computed amount, which is impossible to create when using `BackgroundTasks`. All functional tasks use the *Redis* database as the temporary storage for their returned values, which is the reason there is a backend parameter in the Celery constructor.

Calling the task

FastAPI services can call these tasks using the `apply_async()` or `delay()` function. The latter is the easier option since it is preconfigured and only needs the parameters for the transaction to get the result. The `apply_async()` function is a better option since it accepts more details that can optimize the task execution. These details are `queue`, `time_limit`, `retry`, `ignore_result`, `expires`, and some `kwargs` of arguments. But both these functions return an `AsyncResult` object, which returns resources such as the task's `state`, the `wait()` function to help the task finish its operation, and the `get()` function to return its computed value or an exception. The following code is a coroutine API service that calls the `services.billing.tasks.create_total_payables_year_celery` task using the `apply_async` method:

```
@router.post("/billing/total/payable")
async def compute_payables_yearly(billing_date:date):
    repo = BillingVendorRepository()
    result = await repo.join_vendor_billing()
    total_result = create_total_payables_year_celery
        .apply_async(queue='default',
            args=(billing_date, result))
    total_payable = total_result.get(timeout=1)
    return {"total_payable": total_payable }
```

Setting `task_create_missing_queues` to `True` at the `CeleryConfig` setup is always recommended because it automatically creates the task queue, default or not, once the worker server starts. The worker server places all the loaded tasks in a task queue for execution, monitoring, and result retrieval. Thus, we should always define a task queue in the `apply_async()` function's argument before extracting `AsyncResult`.

The `AsyncResult` object has a `get()` method that releases the returned value of the task from the `AsyncResult` instance, with or without a timeout. In the `compute_payables_yearly()` service, the amount payable in `AsyncResult` is retrieved by the `get()` function with a timeout of 5 seconds. Let us now deploy and run our tasks using the Celery server

Starting the worker server

Running the Celery worker creates a single process that handles and manages all the queued tasks. The worker needs to know in which module the Celery instance is created, together with the tasks to establish the server process. In our prototype, the `services.billing` module is where we place our Celery application. Thus, the complete command to start the worker is the following:

```
celery  -A services.billing worker -Q default -P solo -c 2 -l
info
```

Here, -A specifies the module of our Celery object and tasks. The -Q option indicates that the worker will be using a *low-*, *normal-*, *or high-priority* queue. But first, we need to set `task_create_missing_queues` to `True` in the Celery setup. We also need to indicate the number of threads that the worker needs for task execution by adding the -c option. The -P option specifies the type of *thread pool* that the worker will be utilizing. By default, the Celery worker uses the `prefork pool` applicable to most CPU-bound transactions. Other options are *solo*, *eventlet*, and *gevent*, but our setup will be utilizing *solo*, the most suitable choice for running CPU-intensive tasks in a microservice environment. On the other hand, the -l option enables the logger we set using `get_task_logger()` during the setup. Now, there are also ways to monitor our running tasks and one of those options is to use the Flower tool.

Monitoring the tasks

Flower is Celery's monitoring tool that observes and monitors all tasks executions by generating a real-time audit on a web-based platform. But first, we need to install it using `pip`:

```
pip install flower
```

And then, we run the following `celery` command with the `flower` option:

```
celery -A services.billing flower
```

To view the audit, we run `http://localhost:5555/tasks` on a browser. *Figure 8.3* shows a *Flower* snapshot of an execution log incurred by the `services.billing.tasks.create_total_payables_year_celery` task:

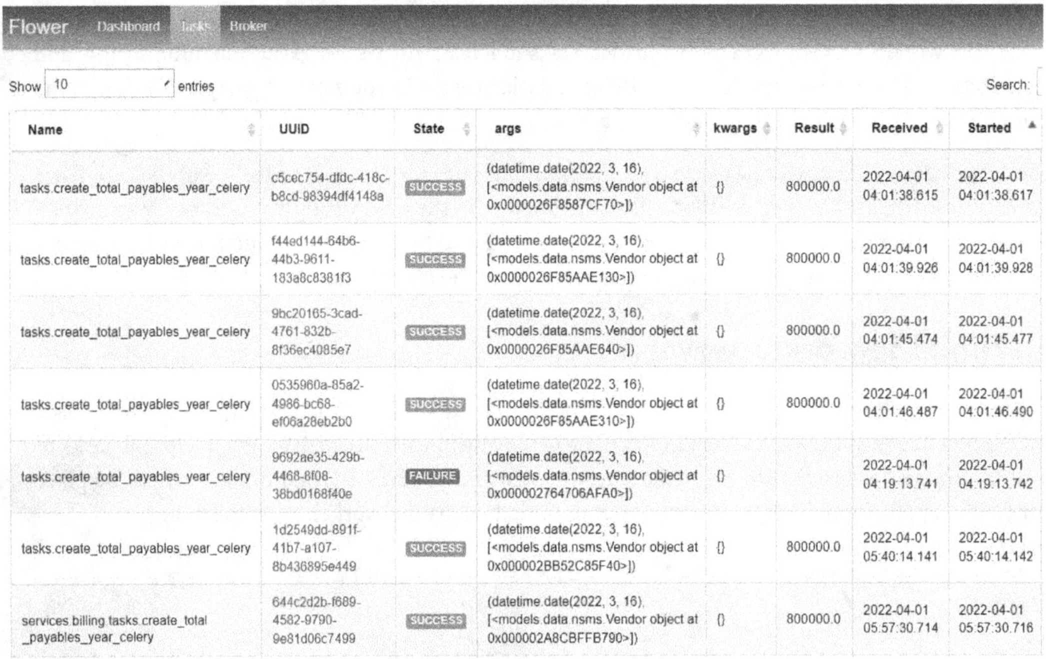

Name	UUID	State	args	kwargs	Result	Received	Started
tasks.create_total_payables_year_celery	c5cec754-dfdc-418c-b8cd-98394df4148a	SUCCESS	(datetime.date(2022, 3, 16), [<models.data.nsms.Vendor object at 0x0000026F8587CF70>])	{}	800000.0	2022-04-01 04:01:38.615	2022-04-01 04:01:38.617
tasks.create_total_payables_year_celery	f44ed144-64b6-44b3-9611-183a8c8381f3	SUCCESS	(datetime.date(2022, 3, 16), [<models.data.nsms.Vendor object at 0x0000026F85AAE130>])	{}	800000.0	2022-04-01 04:01:39.926	2022-04-01 04:01:39.928
tasks.create_total_payables_year_celery	9bc20165-3cad-4761-832b-8f36ec4085e7	SUCCESS	(datetime.date(2022, 3, 16), [<models.data.nsms.Vendor object at 0x0000026F85AAE640>])	{}	800000.0	2022-04-01 04:01:45.474	2022-04-01 04:01:45.477
tasks.create_total_payables_year_celery	0535960a-85a2-4986-bc68-ef06a28eb2b0	SUCCESS	(datetime.date(2022, 3, 16), [<models.data.nsms.Vendor object at 0x0000026F85AAE310>])	{}	800000.0	2022-04-01 04:01.46.487	2022-04-01 04:01:46.490
tasks.create_total_payables_year_celery	9692ae35-429b-4468-8f08-38bd0168f40e	FAILURE	(datetime.date(2022, 3, 16), [<models.data.nsms.Vendor object at 0x000002764706AFA0>])	{}		2022-04-01 04:19:13.741	2022-04-01 04:19:13.742
tasks.create_total_payables_year_celery	1d2549dd-891f-41b7-a107-8b436895e449	SUCCESS	(datetime.date(2022, 3, 16), [<models.data.nsms.Vendor object at 0x000002BB52C85F40>])	{}	800000.0	2022-04-01 05:40:14.141	2022-04-01 05:40:14.142
services.billing.tasks.create_total_payables_year_celery	644c2d2b-f689-4582-9790-9e81d06c7499	SUCCESS	(datetime.date(2022, 3, 16), [<models.data.nsms.Vendor object at 0x000002A8CBFFB790>])	{}	800000.0	2022-04-01 05:57:30.714	2022-04-01 05:57:30.716

Figure 8.3 – The Flower monitoring tool

So far, we have used Redis as our in-memory backend database for task results and a message broker. Let us now use another asynchronous message broker that can replace Redis, *RabbitMQ*.

Building message-driven transactions using RabbitMQ

RabbitMQ is a lightweight asynchronous message broker that supports multiple messaging protocols such as *AMQP*, *STOM*, *WebSocket*, and *MQTT*. It requires *erlang* before it works properly in Windows, Linux, or macOS. Its installer can be downloaded from `https://www.rabbitmq.com/download.html`.

Creating the Celery instance

Instead of using Redis as the broker, RabbitMQ is a better replacement as a message broker that will mediate messages between the client and the Celery worker threads. For multiple tasks, RabbitMQ can command the Celery worker to work on these tasks one at a time. The RabbitMQ broker is good for huge messages and it saves these messages to disk memory.

To start, we need to set up a new Celery instance that will utilize the RabbitMQ message broker using its *guest* account. We will use the AMQP protocol as the mechanism for a producer/consumer type of messaging setup. Here is the snippet that will replace the previous Celery configuration:

```
celery = Celery("services.billing",
    broker='amqp://guest:guest@127.0.0.1:5672',
    result_backend='redis://localhost:6379/0',
    include=["services.billing", "models", "config"])
```

Redis will still be the backend resource, as indicated in Celery's `backend_result`, since it is still simple and easy to control and manage when message traffic increases. Let us now use the RabbitMQ to create and manage message-driven transactions.

Monitoring AMQP messaging

We can configure the RabbitMQ management dashboard to monitor the messages handled by RabbitMQ. After the setup, we can log in to the dashboard using the account details to set the broker. *Figure 8.4* shows a screenshot of RabbitMQ's analytics of a situation where the API services called the `services.billing.tasks.create_total_payables_year_celery` task several times:

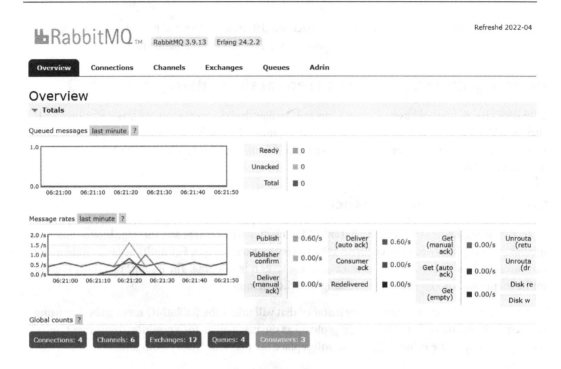

Figure 8.4 – The RabbitMQ management tool

If the RabbitMQ dashboard fails to capture the behavior of the tasks, the *Flower* tool will always be an option for gathering the details about the arguments, kwargs, UUID, state, and processing date of the tasks. And if RabbitMQ is not the right messaging tool, we can always resort to *Apache Kafka*.

Building publish/subscribe messaging using Kafka

As with RabbitMQ, *Apache Kafka* is an asynchronous messaging tool used by applications to send and store messages between producers and consumers. However, it is faster than RabbitMQ because it uses *topics* with partitions where producers can append various types of messages across these minute folder-like structures. In this architecture, the consumers can consume all these messages in a parallel mode, unlike in queue-based messaging, which enables producers to send multiple messages to a queue that can only allow message consumption sequentially. Within this publish/subscribe architecture, Kafka can handle an exchange of large quantities of data per second in continuous and real-time mode.

There are three Python extensions that we can use to integrate the FastAPI services with Kafka, namely the kafka-python, confluent-kafka, and pykafka extensions. Our online *newsstand* prototype will use kafka-python, so we need to install it using the pip command:

```
pip install kafka-python
```

Among the three extensions, it is only with `kafka-python` that we can channel and apply Java API libraries to Python for the implementation of a client. We can download Kafka from `https://kafka.apache.org/downloads`.

Running the Kafka broker and server

Kafka has a *ZooKeeper* server that manages and synchronizes the exchange of messages within Kafka's distributed system. The ZooKeeper server runs as the broker that monitors and maintains the Kafka nodes and topics. The following command starts the server:

```
C:\..\kafka\bin\windows\zookeeper-server-start.bat
C:\..\kafka\config\zookeeper.properties
```

Now, we can start the Kafka server by running the following console command:

```
C:\..\kafka\bin\windows\kafka-server-start.bat
C:\..\kafka\config\server.properties
```

By default, the server will run on localhost at port `9092`.

Creating the topic

When the two servers have started, we can now create a topic called `newstopic` through the following command:

```
C:\..\kafka-topics.bat --create --bootstrap-server
localhost:9092 --replication-factor 1 --partitions 3
--topic newstopic
```

The `newstopic` topic has three (3) partitions that will hold all the appended messages of our FastAPI services. These are also the points where the consumers will simultaneously access all the published messages.

Implementing the publisher

After creating the topic, we can now implement a producer that publishes messages to the Kafka cluster. The `kafka-python` extension has a `KafkaProducer` class that instantiates a single thread-safe producer for all the running FastAPI threads. The following is an API service that sends a newspaper messenger record to the Kafka `newstopic` topic for the consumer to access and process:

```
from kafka import KafkaProducer


producer = KafkaProducer(
```

```
        bootstrap_servers='localhost:9092')


def json_date_serializer(obj):
    if isinstance(obj, (datetime, date)):
        return obj.isoformat()
    raise TypeError ("Data %s not serializable" %
            type(obj))


@router.post("/messenger/kafka/send")
async def send_messnger_details(req: MessengerReq):
    messenger_dict = req.dict(exclude_unset=True)
    producer.send("newstopic",
        bytes(str(json.dumps(messenger_dict,
            default=json_date_serializer)), 'utf-8'))
    return {"content": "messenger details sent"}
```

The coroutine API service, send_messenger_details(), asks for details about a newspaper messenger and stores them in a BaseModel object. And then, it sends the dictionary of profile details to the cluster in byte format. Now, one of the options to consume Kafka tasks is to run its built-in kafka-console-consumer.bat command.

Running a consumer on a console

Running the following command from the console is one way to consume the current messages from the newstopic topic:

```
kafka-console-consumer.bat --bootstrap-server

127.0.0.1:9092 --topic newstopic
```

This command creates a consumer that will connect to the Kafka cluster to read in real time the current messages from newtopic sent by the producer. *Figure 8.5* shows the capture of the consumer while it is running on the console:

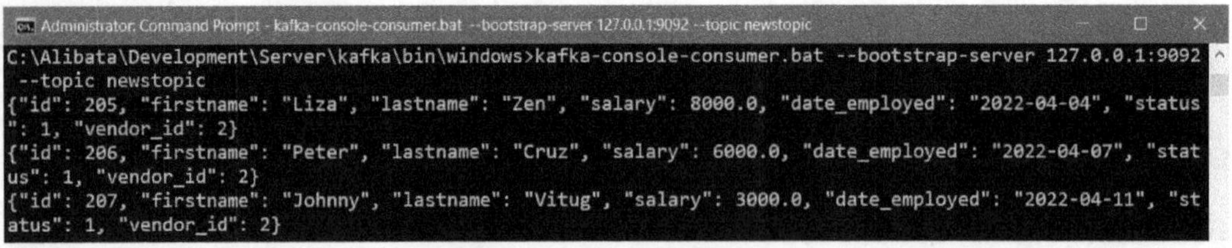

Figure 8.5 – The Kafka consumer

If we want the consumer to read all the messages sent by the producer starting from the point where the Kafka server and broker began running, we need to add the `--from-beginning` option to the command. The following will read all the messages from `newstopic` and continuously capture incoming messages in real time:

```
kafka-console-consumer.bat --bootstrap-server 127.0.0.1:9092
--topic newstopic --from-beginning
```

Another way of implementing a consumer using the FastAPI framework is through SSE. Typical API service implementation will not work with the Kafka consumer requirement since we need a continuously running service that subscribes to `newstopic` for real-time data. So, let us now explore how we create SSE in the FastAPI framework and how it will consume Kafka messages.

Implementing asynchronous Server-Sent Events (SSE)

SSE is a server push mechanism that sends data to the browser without reloading the page. Once subscribed, it generates event-driven streams in real time for various purposes.

Creating SSE in the FastAPI framework only requires the following:

- The `EventSourceResponse` class from the `sse_starlette.see` module
- An event generator

Above all, the framework also allows non-blocking implementation of the whole server push mechanism using coroutines that can run even on *HTTP/2*. The following is a coroutine API service that implements a Kafka consumer using SSE's open and lightweight protocol:

```
from sse_starlette.sse import EventSourceResponse

@router.get('/messenger/sse/add')
async def send_message_stream(request: Request):

    async def event_provider():
        while True:
            if await request.is_disconnected():
                break

            message = consumer.poll()
            if not len(message.items()) == 0:
                for tp, records in message.items():
                    for rec in records:
```

```
            messenger_dict =
             json.loads(rec.value.decode('utf-8'),
              object_hook=date_hook_deserializer )

            repo = MessengerRepository()
            result = await
             repo.insert_messenger(messenger_dict)
            id = uuid4()
            yield {
               "event": "Added … status: {},
                    Received: {}". format(result,
                      datetime.utcfromtimestamp(
                        rec.timestamp // 1000)
                        .strftime("%B %d, %Y
                            [%I:%M:%S %p]")),
               "id": str(id),
               "retry": SSE_RETRY_TIMEOUT,
               "data": rec.value.decode('utf-8')
            }

        await asyncio.sleep(SSE_STREAM_DELAY)
    return EventSourceResponse(event_provider())
```

send_message_stream() is a coroutine API service that implements the whole SSE. It returns a special response generated by an EventSourceResponse function. While the HTTP stream is open, it continuously retrieves data from its source and converts any internal events into SSE signals until the connection is closed.

On the other hand, event generator functions create internal events, which can also be asynchronous. send_message_stream(), for instance, has a nested generator function, event_provider(), which consumes the last message sent by the producer service using the consumer.poll() method. If the message is valid, the generator converts the message retrieved into a dict object and inserts all its details into the database through MessengerRepository. Then, it yields all the internal details for the EventSourceResponse function to convert into SSE signals. *Figure 8.6* shows the data streams generated by send_message_stream() rendered from the browser:

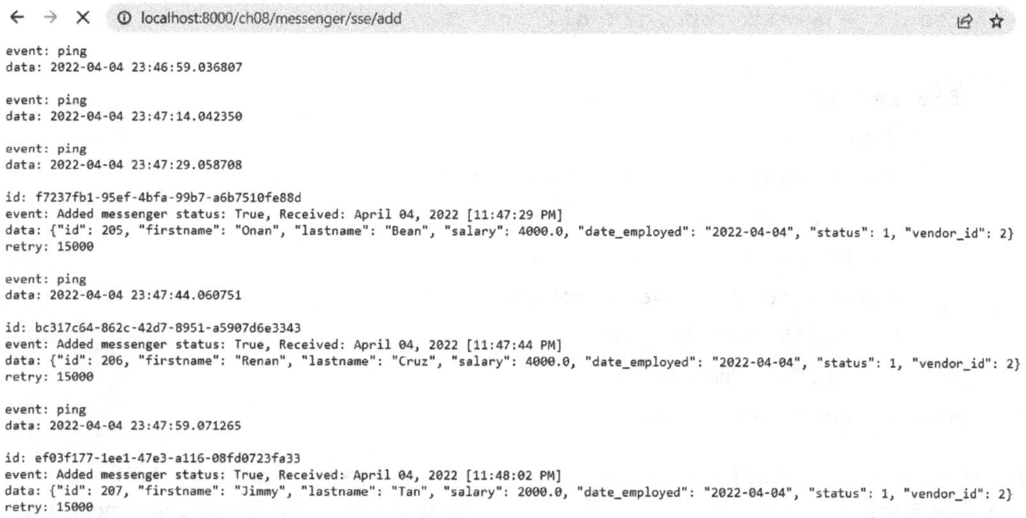

event: ping
data: 2022-04-04 23:46:59.036807

event: ping
data: 2022-04-04 23:47:14.042350

event: ping
data: 2022-04-04 23:47:29.058708

id: f7237fb1-95ef-4bfa-99b7-a6b7510fe88d
event: Added messenger status: True, Received: April 04, 2022 [11:47:29 PM]
data: {"id": 205, "firstname": "Onan", "lastname": "Bean", "salary": 4000.0, "date_employed": "2022-04-04", "status": 1, "vendor_id": 2}
retry: 15000

event: ping
data: 2022-04-04 23:47:44.060751

id: bc317c64-862c-42d7-8951-a5907d6e3343
event: Added messenger status: True, Received: April 04, 2022 [11:47:44 PM]
data: {"id": 206, "firstname": "Renan", "lastname": "Cruz", "salary": 4000.0, "date_employed": "2022-04-04", "status": 1, "vendor_id": 2}
retry: 15000

event: ping
data: 2022-04-04 23:47:59.071265

id: ef03f177-1ee1-47e3-a116-08fd0723fa33
event: Added messenger status: True, Received: April 04, 2022 [11:48:02 PM]
data: {"id": 207, "firstname": "Jimmy", "lastname": "Tan", "salary": 2000.0, "date_employed": "2022-04-04", "status": 1, "vendor_id": 2}
retry: 15000

Figure 8.6 – The SSE data streams

Another way to implement a Kafka consumer is through *WebSocket*. But this time, we will focus on the general procedure of how to create an asynchronous WebSocket application using the FastAPI framework.

Building an asynchronous WebSocket

Unlike in SSE, connection in WebSocket is always *bi-directional*, which means the server and client communicate with each other using a long TCP socket connection. The communication is always in real time and it doesn't require the client or the server to reply to every event sent.

Implementing the asynchronous WebSocket endpoint

The FastAPI framework allows the implementation of an asynchronous WebSocket that can also run on the *HTTP/2* protocol. The following is an example of an asynchronous WebSocket created using the coroutine block:

```
import asyncio
from fastapi import WebSocket


@router.websocket("/customer/list/ws")
async def customer_list_ws(websocket: WebSocket):
    await websocket.accept()
    repo = CustomerRepository()
```

```
    result = await repo.get_all_customer()

    for rec in result:
        data = rec.to_dict()
        await websocket.send_json(json.dumps(data,
            default=json_date_serializer))
        await asyncio.sleep(0.01)
        client_resp = await websocket.receive_json()
        print("Acknowledging receipt of record id
            {}.".format(client_resp['rec_id']))
    await websocket.close()
```

First, we decorate a coroutine function with `@router.websocket()` when using APIRouter, or `@api.websocket()` when using the FastAPI decorator to declare a WebSocket component. The decorator must also define a unique endpoint URL for the WebSocket. Then, the WebSocket function must have an injected `WebSocket` as its first method argument. It can also include other parameters such as query and header parameters.

The `WebSocket` injectable has four ways for sending messages, namely `send()`, `send_text()`, `send_json()`, and `send_bytes()`. Applying `send()` will always manage every message as plain text by default. The previous `customer_list_ws()` coroutine is a WebSocket that sends every customer record in JSON format.

On the other hand, there are also four methods the WebSocket injectable can provide, and these are the `receive()`, `receive_text()`, `receive_json()`, and `receive_bytes()` methods. The `receive()` method expects the message to be in plain-text format by default. Now, our `customer_list_ws()` endpoint expects a JSON reply from a client because it invokes the `receive_json()` method after its send message operation.

The WebSocket endpoint must close the connection right after its transaction is done.

Implementing the WebSocket client

There are many ways to create a WebSocket client but this chapter will focus on utilizing a coroutine API service that will perform a handshake with the asynchronous `customer_list_ws()` endpoint once called on a browser or a `curl` command. Here is the code of our WebSocket client implemented using the `websockets` library that runs on top of the `asyncio` framework:

```
import websockets


@router.get("/customer/wsclient/list/")
async def customer_list_ws_client():
```

```
uri = "ws://localhost:8000/ch08/customer/list/ws"
async with websockets.connect(uri) as websocket:
    while True:
        try:
            res = await websocket.recv()
            data_json = json.loads(res,
                object_hook=date_hook_deserializer)

            print("Received record:
                    {}.".format(data_json))

            data_dict = json.loads(data_json)
            client_resp = {"rec_id": data_dict['id'] }
            await websocket.send(json.dumps(client_resp))

        except websockets.ConnectionClosed:
            break
    return {"message": "done"}
```

After a successful handshake is created by the `websockets.connect()` method, `customer_list_ws_client()` will have a loop running continuously to fetch all incoming consumer details from the WebSocket endpoint. The message received will be converted into its dictionary needed by other processes. Now, our client also sends an acknowledgment notification message back to the WebSocket coroutine with JSON data containing the *customer ID* of the profile. The loop will stop once the WebSocket endpoint closes its connection.

Let us now explore other asynchronous programming features that can work with the FastAPI framework.

Applying reactive programming in tasks

Reactive programming is a paradigm that involves the generation of streams that undergo a series of operations to propagate some changes during the process. Python has an *RxPY* library that offers several methods that we can apply to these streams asynchronously to extract the terminal result as desired by the subscribers.

In the reactive programming paradigm, all intermediate operators working along the streams will execute to propagate some changes if there is an `Observable` instance beforehand and an `Observer` that subscribes to this instance. The main goal of this paradigm is to achieve the desired result at the end of the propagation process using functional programming.

Creating the Observable data using coroutines

It all starts with the implementation of a coroutine function that will emit these streams of data based on a business process. The following is an `Observable` function that emits publication details in `str` format for those publications that did well in sales:

```python
import asyncio
from rx.disposable import Disposable

async def process_list(observer):
        repo = SalesRepository()
        result = await repo.get_all_sales()

        for item in result:
            record = " ".join([str(item.publication_id),
                str(item.copies_issued), str(item.date_issued),
                str(item.revenue), str(item.profit),
                str(item.copies_sold)])
            cost = item.copies_issued * 5.0
            projected_profit = cost - item.revenue
            diff_err = projected_profit - item.profit
            if (diff_err <= 0):
                observer.on_next(record)
            else:
                observer.on_error(record)
        observer.on_completed()
```

An `Observable` function can be synchronous or asynchronous. Our target is to create an asynchronous one such as `process_list()`. The coroutine function should have the following callback methods to qualify as an `Observable` function:

- An `on_next()` method that emits items given a certain condition

- An `on_completed()` method that is executed once when the function has completed the operation

- An `on_error()` method that is called when an error occurs on `Observable`

Our `process_list()` emits the details of the publication that gained some profit. Then, we create an `asyncio` task for the call of the `process_list()` coroutine. We created a nested function, `evaluate_profit()`, which returns the `Disposable` task required by RxPY's `create()` method for the production of the `Observable` stream. The cancellation of this task happens when the `Observable` stream is all consumed. The following is the complete implementation for the execution of the asynchronous `Observable` function and the use of the `create()` method to generate streams of data from this `Observable` function:

```
def create_observable(loop):
    def evaluate_profit(observer, scheduler):
        task = asyncio.ensure_future(
            process_list(observer), loop=loop)
        return Disposable(lambda: task.cancel())
    return rx.create(evaluate_profit)
```

The subscriber created by `create_observable()` is our application's `list_sales_by_quota()` API service. It needs to get the current event loop running for the method to generate the observable. Afterward, it invokes the `subscribe()` method to send a subscription to the stream and extract the needed result. The Observable's `subscribe()` method is invoked for a client to subscribe to the stream and observe the occurring propagations:

```
@router.get("/sales/list/quota")
async def list_sales_by_quota():
    loop = asyncio.get_event_loop()
    observer = create_observable(loop)

    observer.subscribe(
        on_next=lambda value: print("Received Instruction
            to buy {0}".format(value)),
        on_completed=lambda: print("Completed trades"),
        on_error=lambda e: print(e),
        scheduler = AsyncIOScheduler(loop)
    )
    return {"message": "Notification
        sent to the background"}
```

The `list_sales_by_quote()` coroutine service shows us how to subscribe to an Observable. A subscriber should utilize the following callback methods:

- An `on_next()` method to consume all the items from the stream
- An `on_completed()` method to indicate the end of the subscription
- An `on_error()` method to flag an error during the subscription process

And since the `Observable` processes run asynchronously, the scheduler is an optional argument that provides the right manager to schedule and run these processes. The API service used `AsyncIOScheduler` as the appropriate schedule for the subscription. But there are other shortcuts to generating Observables that do not use a custom function.

Creating background process

As when we create continuously running Observables, we use the `interval()` function instead of using a custom `Observable` function. Some observables are designed to end successfully, but some are created to run continuously in the background. The following Observable runs in the background periodically to provide some updates on the total amount received from newspaper subscriptions:

```
import asyncio
import rx
import rx.operators as ops


async def compute_subscriptions():
    total = 0.0
    repo = SubscriptionCustomerRepository()
    result = await repo.join_customer_subscription_total()

    for customer in result:
        subscription = customer.children
        for item in subscription:
            total = total + (item.price * item.qty)
    await asyncio.sleep(1)
    return total


def fetch_records(rate, loop) -> rx.Observable:
    return rx.interval(rate).pipe(
        ops.map(lambda i: rx.from_future(
            loop.create_task(compute_subscriptions()))),
```

```
        ops.merge_all()
    )
```

The `interval()` method creates a stream of data periodically in seconds. But this Observable imposes some propagations on its stream because of the execution of the `pipe()` method. The Observable's `pipe()` method creates a pipeline of reactive operators called the intermediate operators. This pipeline can consist of a chain of operators running one at a time to change items from the streams. It seems that this series of operations creates multiple subscriptions on the subscriber. So, `fetch_records()` has a `map()` operator in its pipeline to extract the result from the `compute_subcription()` method. It uses `merge_all()` at the end of the pipeline to merge and flatten all substreams created into one final stream, the stream expected by the subscriber. Now, we can also generate Observable data from files or API response.

Accessing API resources

Another way of creating an Observable is using the `from_()` method, which extracts resources from files, databases, or API endpoints. The Observable function retrieves its data from a JSON document generated by an API endpoint from our application. The assumption is that we are running the application using `hypercorn`, which uses *HTTP/2*, and so we need to bypass the TLS certificate by setting the `verify` parameter of `httpx.AsyncClient()` to `False`.

The following code highlights the `from_()` in the `fetch_subscription()` operation, which creates an Observable that emits streams of `str` data from the `https://localhost:8000/ch08/subscription/list/all` endpoint. These reactive operators of the Observable, namely `filter()`, `map()`, and `merge_all()`, are used to propagate the needed contexts along the stream:

```
async def fetch_subscription(min_date:date,
        max_date:date, loop) -> rx.Observable:
    headers = {
            "Accept": "application/json",
            "Content-Type": "application/json"
        }
    async with httpx.AsyncClient(http2=True,
            verify=False) as client:
        content = await
            client.get('https://localhost:8000/ch08/
            subscription/list/all', headers=headers)
    y = json.loads(content.text)
    source = rx.from_(y)
    observable = source.pipe(
        ops.filter(lambda c: filter_within_dates(
```

```
                    c, min_date, max_date)),
        ops.map(lambda a: rx.from_future(loop.create_task(
            convert_str(a)))),
        ops.merge_all(),
    )
    return observable
```

The `filter()` method is another pipeline operator that returns Boolean values from a validation rule. It executes the following `filter_within_dates()` to verify whether the record retrieved from the JSON document is within the date range specified by the subscriber:

```
def filter_within_dates(rec, min_date:date, max_date:date):
    date_pur = datetime.strptime(
            rec['date_purchased'], '%Y-%m-%d')
    if date_pur.date() >= min_date and
            date_pur.date() <= max_date:
        return True
    else:
        return False
```

On the other hand, the following `convert_str()` is a coroutine function executed by the `map()` operator to generate a concise profile detail of the newspaper subscribers derived from the JSON data:

```
async def convert_str(rec):
    if not rec == None:
        total = rec['qty'] * rec['price']
        record = " ".join([rec['branch'],
            str(total), rec['date_purchased']])
        await asyncio.sleep(1)
        return record
```

Running these two functions modifies the original emitted data stream from JSON to a date-filtered stream of `str` data. The coroutine `list_dated_subscription()` API service, on the other hand, subscribes to `fetch_subscription()` to extract the newspaper subscriptions within the `min_date` and `max_date` range:

```
@router.post("/subscription/dated")
async def list_dated_subscription(min_date:date,
            max_date:date):
```

```
loop = asyncio.get_event_loop()
observable = await fetch_subscription(min_date,
        max_date, loop)

observable.subscribe(
   on_next=lambda item:
      print("Subscription details: {}.".format(item)),
   scheduler=AsyncIOScheduler(loop)
)
```

Although the FastAPI framework does not yet fully support reactive programming, we can still create coroutines that can work with various RxPY utilities. Now, we will explore how coroutines are not only for background processes but also for FastAPI event handlers.

Customizing events

The FastAPI framework has special functions called *event handlers* that execute before the application starts up and during shutdown. These events are activated every time the uvicorn or hypercorn server reloads. Event handlers can also be coroutines.

Defining the startup event

The *startup event* is an event handler that the server executes when it starts up. We decorate the function with the @app.on_event("startup") decorator to create a startup event. Applications may require a startup event to centralize some transactions, such as the initial configuration of some components or the set up of data-related resources. The following example is the application startup event that opens a database connection for the GINO repository transactions:

```
app = FastAPI()

@app.on_event("startup")
async def initialize():
    engine = await db.set_bind("postgresql+asyncpg://
        postgres:admin2255@localhost:5433/nsms")
```

This initialize() event is defined in our application's main.py file so that GINO can only create the connection once every server reload or restart.

Defining shutdown events

Meanwhile, the *shutdown event* cleans up unwanted memory, destroys unwanted connections, and logs the reason for shutting down the application. The following is the shutdown event of our application that closes the GINO database connection:

```
@app.on_event("shutdown")
async def destroy():
    engine, db.bind = db.bind, None
    await engine.close()
```

We can define startup and shutdown events in APIRouter but be sure this will not cause transaction overlapping or collision with other routers. Moreover, event handlers do not work in mounted sub-applications.

Summary

The use of coroutines is one of the factors that makes the FastAPI microservice application fast, aside from its use of an ASGI-based server. This chapter has proven that using coroutines to implement API services will improve the performance better than utilizing more threads in the thread pool. Since the framework runs on an asyncio platform, we can utilize asyncio utilities to design various design patterns to manage the CPU-bound and I/O-bound services.

This chapter used Celery and Redis for creating and managing asynchronous background tasks for behind-the-scenes transactions such as logging, system monitoring, time-sliced computations, and batch jobs. We learned that RabbitMQ and Apache Kafka provided an integrated solution for building asynchronous and loosely coupled communication between FastAPI components, especially for the message-passing part of these interactions. Most importantly, coroutines were applied to create these asynchronous and non-blocking background processes and message-passing solutions to enhance performance. Reactive programming was also introduced in this chapter through the RxPy extension module.

This chapter, in general, concludes that the FastAPI framework is ready to build a microservice application that has a *reliable*, *asynchronous*, *message-driven*, *real-time message-passing*, and *distributed core system*. The next chapter will highlight other FastAPI features that provide integrations with UI-related tools and frameworks, API documentation using OpenAPI Specification, session handling, and circumventing CORS.

Part 3: Infrastructure-Related Issues, Numerical and Symbolic Computations, and Testing Microservices

In this final part of the book, we will discuss other essential microservice features, such as distributed tracing and logging, service registries, virtual environments, and API metrics. Serverless deployment using Docker and Docker Compose with NGINX as a reverse proxy will also be covered. Furthermore, we will look at FastAPI as a framework for building scientific applications using numerical algorithms from the `numpy`, `scipy`, `sympy`, and `pandas` modules to model, analyze, and visualize the mathematical and statistical solutions of its API services.

This part comprises the following chapters:

- *Chapter 9, Utilizing Other Advanced Features*
- *Chapter 10, Solving Numerical, Symbolic, and Graphical Problems*
- *Chapter 11, Adding Other Microservice Features*

9

Utilizing Other Advanced Features

The previous chapters have already showcased several essential core features of the FastAPI framework. However, there are features not truly inherent to the framework that can help fine-tune performance and patch missing links in our implementations. These include session handling, managing **Cross-Origin Resource Sharing** (**CORS**)-related issues, and selecting the appropriate rendition types for an application.

Aside from the built-in features, there are workaround solutions proven to work with FastAPI when applied to the application, such as its session handling mechanism, which can function well using a **JWT**, and `SessionMiddleware`. Concerning middleware, this chapter will also explore ways of customizing request and response filters other than applying the `@app.middleware` decorator. Other issues such as using custom `APIRoute` and `Request` will be covered in this chapter to guide us on managing incoming *byte body*, *form*, or *JSON* data. Moreover, this chapter will highlight how to test FastAPI components using the `pytest` framework and the `fastapi.testclient` library and how we can document the endpoint using the *OpenAPI 3.x* specification.

Overall, the main objective of this chapter is to provide us with other solutions that can help us complete our microservice applications. In this chapter, the following topics are included:

- Applying session management
- Managing the CORS mechanism
- Customizing `APIRoute` and `Request`
- Choosing appropriate responses
- Applying the OpenAPI 3.x specification
- Testing the API endpoints

Technical requirements

Although not data analysis-related, our application prototype for this chapter is the **online restaurant review system**, which will gather ordinal and nominal ratings and feedback for restaurants. The software aims to gather rates and feedback to establish the user profiles of restaurants and conduct surveys concerning their food menus, facilities, ambiance, and services. The prototype will use MongoDB as the data storage and asynchronous ODMantic as its ORM. All of the code is uploaded to https://github.com/PacktPublishing/Building-Python-Microservices-with-FastAPI under the ch09 project.

Applying session management

Session management is a feature used for managing requests and responses created by a user's access to an application. It is also about creating and sharing data across a user session. Many frameworks usually include session handling features in their security plugins but not FastAPI. Creating user sessions and storing session data are two separate programming concerns in FastAPI. We use a JWT to establish a user session and Starlette's `SessionMiddleware` to create and retrieve session data. Creating user sessions and storing session data are two entirely different programming solutions in FastAPI. We use JWT to establish a user session and Starlette's `SessionMiddleware` to create and retrieve session data.

Creating user sessions

We have already proven the importance of JWT when it comes to securing FastAPI microservice applications in *Chapter 7, Securing the REST APIs*. However, here, the JWT is applied to create a session based on user credentials. In the `api/login.py` router, an `authenticate()` API service is implemented to create a user session for an authenticated user. It is inherent for FastAPI to generate user sessions utilizing the browser cookies. The following snippet shows the authentication process that uses the cookie values:

```
from util.auth_session import secret_key
from jose import jwt


@router.post("/login/authenticate")
async def authenticate(username:str, password: str,
    response: Response, engine=Depends(create_db_engine)):
    repo:LoginRepository = LoginRepository(engine)
    login = await repo.get_login_username(username,
                        password)
    if login == None:
            raise HTTPException(
```

```
                        status_code=status.HTTP_403_FORBIDDEN,
                        detail="Invalid authentication"
                )
    token = jwt.encode({"sub": username}, secret_key)
    response.set_cookie("session", token)
    return {"username": username}
```

The service will verify through `LoginRepository` whether the user is a valid account using its `username` and `password` credentials. If the user is a certified one, it will use a JWT to create a token derived from a certain `secret_key` generated using the following command:

```
openssl rand -hex 32
```

The token key will serve as the session ID of the cookie-based session. With the `username` credential as its payload, the JWT will be stored as a browser cookie named `session`.

To ensure that `session` has been applied, all subsequent requests must undergo authentication by the cookie-based session through the `APIKeyCookie` class, an API class of the `fastapi.security` module that implements cookie-based authentication. The `APIKeyCookie` class fetches the session before it is injected into a dependable function for the JWT decoding through the `secret_key` value used to generate the session ID. The following dependable function in `util/auth_session.py` will verify every access to each endpoint of the application:

```
from fastapi.security import APIKeyCookie
from jose import jwt


cookie_sec = APIKeyCookie(name="session")
secret_key = "pdCFmblRt4HWKNpWkl52Jnq3emH3zzg4b80f+4AFVC8="


async def get_current_user(session: str =
    Depends(cookie_sec), engine=Depends(create_db_engine)):
    try:
        payload = jwt.decode(session, secret_key)
        repo:LoginRepository = LoginRepository(engine)
        login = await repo.validate_login(
                    payload["sub"])
        if login == None:
            raise HTTPException(
                status_code=status.HTTP_403_FORBIDDEN,
                detail="Invalid authentication"
```

```
            )
        else:
            return login
    except Exception:
        raise HTTPException(
            status_code=status.HTTP_403_FORBIDDEN,
            detail="Invalid authentication"
        )
```

The preceding function is injected into every API endpoint to impose user session verification. When an endpoint is requested, this function will decode the token and extract the username credential for account validation. Then, it will issue *Status Code 403 (Forbidden)* if the user is an *unauthenticated* one or the *session is not valid*. An example of an authenticated service can be found in the following implementation:

```
from util.auth_session import import get_current_user

@router.post("/restaurant/add")
async def add_restaurant(req:RestaurantReq,
        engine=Depends(create_db_engine),
        user: str = Depends(get_current_user)):
    restaurant_dict = req.dict(exclude_unset=True)
    restaurant_json = dumps(restaurant_dict,
            default=json_datetime_serializer)
    repo:RestaurantRepository =
            RestaurantRepository(engine)
    result = await repo.insert_restaurant(
            loads(restaurant_json))
    if result == True:
        return req
    else:
        return JSONResponse(content={"message":
        "insert login unsuccessful"}, status_code=500)
```

The add_restaurant() service is an endpoint that adds a restaurant Document to the MongoDB collection. But before the transaction proceeds, it checks first whether there is a cookie-based session through the injected get_current_user() dependable function.

Managing session data

Unfortunately, adding and retrieving session data is not part of `APIKeyCookie`-based session authentication. The JWT payload must only include the username but not all credentials and body of data. To manage session data, we need to create a separate session using Starlette's `SessionMiddleware`. Although FastAPI has its `fastapi.middleware` module, it still supports Starlette's built-in middleware.

We mentioned middleware in *Chapter 2, Exploring the Core Features*, and showed its implementation using the `@app.middleware` decorator. And we have proven that it acts as a filter for all incoming requests and outgoing responses to the services. This time, we will not custom implement a middleware but built-in middleware classes.

Middleware is implemented, configured, and activated in the `main.py` module where the instance of `FastAPI` is located because `APIRouter` cannot add middleware. We enable the `middleware` parameter of the FastAPI constructor and add to that List-type parameter the built-in `SessionMiddleware` with its `secret_key` and the name of the new session as constructor parameters using the injectable class, `Middleware`. The following code snippet of `main.py` shows you how to configure this:

```
from starlette.middleware.sessions import SessionMiddleware


app = FastAPI(middleware=[
        Middleware(SessionMiddleware,
        secret_key=
            '7UzGQS7woBazLUtVQJG39ywOP7J71kPkB0UmDhMgBR8=',
        session_cookie="session_vars")])
```

Another way of adding middleware is to utilize the `add_middleware()` function of the `FastAPI` decorator. Initially, adding `SessionMiddleware` will create another cookie-based session that will handle *session-scoped data*. It is the only way since there is no direct support from FastAPI regarding session handling mechanisms where a user session is created not only for security but also for handling session objects.

To add session data to our newly created session, `session_vars`, we need to inject `Request` into each endpoint service and utilize its session dictionary to store the session-scoped objects. The following `list_restaurants()` service retrieves the list of restaurants from the database, extracts all the restaurant names, and shares the list of names across the session through `request.session[]`:

```
@router.get("/restaurant/list/all")
async def list_restaurants(request: Request,
        engine=Depends(create_db_engine),
        user: str = Depends(get_current_user)):
```

```
            repo:RestaurantRepository =
                    RestaurantRepository(engine)
            result = await repo.get_all_restaurant()
            resto_names = [resto.name for resto in result]
            request.session['resto_names'] = resto_names
            return result


@router.get("/restaurant/list/names")
async def list_restaurant_names(request: Request,
            user: str = Depends(get_current_user)):
        resto_names = request.session['resto_names']
        return resto_names
```

On the other hand, the `list_restaurant_names()` service retrieves the `resto_names` session data through `request.session[]` and returns it as its response. By the way, it is due to `SessionMiddleware` that `session[]` exists. Otherwise, the use of this dictionary will raise an change to - exception.

Removing the sessions

It is always mandatory to log out from the application when done with the transactions to remove all the sessions created. Since the easiest and most direct way of creating sessions is through browser cookies, removing all the sessions protects the application from any compromise. The following `/ch09/logout` endpoint removes our sessions, `session` and `session_vars`, which technically logs out the user from the application:

```
@router.get("/logout")
async def logout(response: Response,
            user: str = Depends(get_current_user)):
    response.delete_cookie("session")
    response.delete_cookie("session_vars")
    return {"ok": True}
```

The `delete_cookie()` method of the `Response` class removes any existing browser session utilized by the application.

Customizing BaseHTTPMiddleware

The default approach in managing FastAPI sessions is through cookies, and it does not offer any other options such as database-backed, cached, and file-based sessions. The best way to implement non-cookie-based strategies for managing user sessions and session data is to customize `BaseHTTPMiddleware`. The following custom middleware is a prototype that creates user sessions for authenticated users:

```python
from repository.login import LoginRepository
from repository.session import DbSessionRepository
from starlette.middleware.base import BaseHTTPMiddleware
from datetime import date, datetime
import re

from odmantic import AIOEngine
from motor.motor_asyncio import AsyncIOMotorClient

class SessionDbMiddleware(BaseHTTPMiddleware):
    def __init__(self, app, sess_key: str,
                     sess_name:str, expiry:str):
        super().__init__(app)
        self.sess_key = sess_key
        self.sess_name = sess_name
        self.expiry = expiry
        self.client_od =
         AsyncIOMotorClient(f"mongodb://localhost:27017/")
        self.engine =
         AIOEngine(motor_client=self.client_od,
            database="orrs")

    async def dispatch(self, request: Request, call_next):
        try:
            if re.search(r'\bauthenticate\b',
                    request.url.path):
                credentials = request.query_params
                username = credentials['username']
                password = credentials['password']
                repo_login:LoginRepository =
```

```
                    LoginRepository(self.engine)
        repo_session:DbSessionRepository =
                DbSessionRepository(self.engine)

        login = await repo_login.
            get_login_credentials(username, password)

        if login == None:
            self.client_od.close()
            return JSONResponse(status_code=403)
        else:
            token = jwt.encode({"sub": username},
                self.sess_key)
            sess_record = dict()
            sess_record['session_key'] =
                self.sess_key
            sess_record['session_name'] =
                self.sess_name
            sess_record['token'] = token
            sess_record['expiry_date'] =
                datetime.strptime(self.expiry,
                    '%Y-%m-%d')
            await repo_session.
                insert_session(sess_record)
            self.client_od.close()
            response = await call_next(request)
            return response
    else:
        response = await call_next(request)
        return response
except Exception as e :
    return JSONResponse(status_code=403)
```

As discussed in *Chapter 2, Exploring the Core Features*, **middleware** is a low-level implementation of a filter for all requests and responses of the applications. So, first, `SessionDbMiddleware` will filter our `/ch09/login/authenticate` endpoint for the `username` and `password` query parameters, check whether the user is a registered one, and generate a database-backed session from the JWT. Afterward, endpoints can validate all their requests from the session stored in the database. The `/ch09/logout` endpoint will not include the deletion of the session from the database using its repository transactions, as shown in the following code:

```
@router.get("/logout")
async def logout(response: Response,
        engine=Depends(create_db_engine),
        user: str = Depends(get_current_user)):
    repo_session:DbSessionRepository =
            DbSessionRepository(engine)
    await repo_session.delete_session("session_db")
    return {"ok": True}
```

Note that `DbSessionRepository` is a custom repository implementation for our prototype, and it has a `delete_session()` method that will remove the session through its name from the `db_session` collection of our MongoDB database.

Another type of middleware that can help FastAPI applications resolve issues regarding the CORS browser mechanism is `CORSMiddleware`.

Managing the CORS mechanism

When integrating API endpoints with various frontend frameworks, we often encounter the *"no 'access-control-allow-origin' header present"* error from our browser. Nowadays, this setup is an HTTP-header-based mechanism of any browser, which requires the backend server to provide the browser with the "origin" details of the server-side application, which includes the server domain, scheme, and port. This mechanism is called CORS, which happens when the frontend application and its web resources belong to a different domain area than the backend app. Nowadays, browsers prohibit cross-origin requests between the server-side and frontend applications for security reasons.

To resolve this issue, we need our `main.py` module to place all the origins of our application and other integrated resources used by the prototype inside a `List`. Then, we import the built-in `CORSMiddleware` from the `fastapi.middleware.cors` module and add that to the `FastAPI` constructor with the list of origins, which should not be too long to avoid overhead from validating each URL. The following code snippet shows the injection of `CORSMiddleware` into the `FastAPI` constructor:

```python
origins = [
    "https://192.168.10.2",
    "http://192.168.10.2",
    "https://localhost:8080",
    "http://localhost:8080"
]
app = FastAPI(middleware=[
            Middleware(SessionMiddleware, secret_key=
              '7UzGQS7woBazLUtVQJG39ywOP7J7lkPkB0UmDhMgBR8=',
                session_cookie="session_vars"),
            Middleware(SessionDbMiddleware, sess_key=
              '7UzGQS7woBazLUtVQJG39ywOP7J7lkPkB0UmDhMgBR8=',
                sess_name='session_db', expiry='2020-10-10')
            ])
app.add_middleware(CORSMiddleware, max_age=3600,
    allow_origins=origins, allow_credentials=True,
    allow_methods= ["POST", "GET", "DELETE",
      "PATCH", "PUT"], allow_headers=[
        "Access-Control-Allow-Origin",
        "Access-Control-Allow-Credentials",
        "Access-Control-Allow-Headers",
        "Access-Control-Max-Age"])
```

This time, we used FastAPI's `add_middleware()` function to add CORS support to our application. Aside from `allow_origins`, we also need to add into `CORSMiddleware` the `allow_credentials` parameter, which adds `Access-Control-Allow-Credentials: true` to the response header for the browser to recognize the domain origin matches and send an `Authorization` cookie to allow the request. Also, we must include the `allow_headers` parameter, which registers a list of acceptable header keys during browser interaction. Aside from `Accept`, `Accept-Language`, `Content-Language`, and `Content-Type`, which are included by default, we need to register `Access-Control-Allow-Origin`, `Access-Control-Allow-Credentials`, `Access-Control-Allow-Headers`, and `Access-Control-Max-Age` explicitly instead of using the asterisk (`*`). The `allow_headers` parameter must also be part of the middleware to specify other HTTP methods that need to be supported by the browser. And lastly, the `max_age` parameter must also be in the configuration because we need to tell the browser the amount of time it will cache all the resources loaded into the browser.

If the application needs additional CORS support features, customizing the `CORSMiddleware` to extend some built-in utilities and features to manage CORS is a better solution.

By the way, it is not only the middleware that we can subclass and use to create custom implementations of but also the `Request` data and API routes.

Customizing APIRoute and Request

Middleware can process incoming `Request` data and outgoing `Response` objects of all API methods in a FastAPI application, except that it cannot manipulate the message body, attach state objects from the `Request` data, or modify the response object before the client consumes it. Only `APIRoute` and `Request` customization can give us a full grasp of how to control the request and response transaction. The control might include determining whether the incoming data is a byte body, form, or JSON and providing an effective logging mechanism, exception handling, content transformation, and extraction.

Managing body, form, or JSON data

Unlike in middleware, customizing `APIRoute` does not apply to all the API endpoints. Implementing `APIRoute` for some `APIRouter` will only impose new routing rules to those affected endpoints, while the other services can pursue the default request and response process. For instance, the following customization is responsible for data extraction that only applies to the endpoints of `api.route_extract.router`:

```
from fastapi.routing import APIRoute
from typing import Callable
from fastapi import Request, Response


class ExtractContentRoute(APIRoute):
    def get_route_handler(self) -> Callable:
        original_route_handler =
                super().get_route_handler()


        async def custom_route_handler(request: Request)
                    -> Response:
            request = ExtractionRequest(request.scope,
```

```
                request.receive)
        response: Response = await
                original_route_handler(request)
        return response
    return custom_route_handler
```

Customizing APIRoute requires the creation of a Python **closure** that will directly manage the
Request and Response flow from APIRoute's original_route_handler. On the other
hand, our ExtractContentRoute filter uses a custom ExtractionRequest that identifies
and processes each type of incoming request data separately. The following is the implementation of
ExtractionRequest that will replace the default Request object:

```python
class ExtractionRequest(Request):
    async def body(self):
        body = await super().body()
        data = ast.literal_eval(body.decode('utf-8'))
        if isinstance(data, list):
            sum = 0
            for rate in data:
                sum += rate
            average = sum / len(data)
            self.state.sum = sum
            self.state.avg = average
        return body

    async def form(self):
        body = await super().form()
        user_details = dict()
        user_details['fname'] = body['firstname']
        user_details['lname'] = body['lastname']
        user_details['age'] = body['age']
        user_details['bday'] = body['birthday']
        self.session["user_details"] = user_details
        return body

    async def json(self):
        body = await super().json()
```

```
        if isinstance(body, dict):

            sum = 0
            for rate in body.values():
                sum += rate

            average = sum / len(body.values())
            self.state.sum = sum
            self.state.avg = average
        return body
```

To activate this `ExtractionRequest`, we need to set the `route_class` of the `APIRouter` of the endpoints to `ExtractContentRoute`, as shown in the following snippet:

```
router = APIRouter()
router.route_class = ExtractContentRoute
```

There are three methods of choice to override when managing various request bodies:

- `body()`: This manages incoming request data that is in bytes
- `form()`: This processes incoming form data
- `json()`: This manages incoming parsed JSON data
- `stream()`: This accesses the body via a chunk of bytes using the `async for` construct

All of these methods return the original request body in bytes back to the service.

In `ExtractionRequest`, we have implemented three interface methods from the given choices to filter and process all incoming requests of the API endpoints defined in the `/api/route_extract.py` module.

The following `create_profile()` service accepts profile data from the client and implements the `ExtractContentRoute` filter, which will store all of this profile data in the dictionary using session handling:

```
@router.post("/user/profile")
async def create_profile(req: Request,
        firstname: str = Form(...),
        lastname: str = Form(...), age: int = Form(...),
        birthday: date = Form(...),
        user: str = Depends(get_current_user)):
```

```
    user_details = req.session["user_details"]
    return {'profile' : user_details}
```

The overridden `form()` method of `ExtractionRequest` is responsible for the `user_details` attribute containing all the user details.

On the other hand, the given `set_ratings()` method has an incoming dictionary of various ratings in which the `json()` override will derive some basic statistics. All the results will be returned as `Request`'s state objects or request attributes:

```
@router.post("/rating/top/three")
async def set_ratings(req: Request, data :
 Dict[str, float], user: str = Depends(get_current_user)):
    stats = dict()
    stats['sum'] = req.state.sum
    stats['average'] = req.state.avg
    return {'stats' : stats }
```

And lastly, the preceding `compute_data()` service will have an incoming list of ratings as a source of some basic statistics like in the previous service. The `body()` method override of `ExtractionRequest` will process the computation:

```
@router.post("/rating/data/list")
async def compute_data(req: Request, data: List[float],
   user: str = Depends(get_current_user)):
    stats = dict()
    stats['sum'] = req.state.sum
    stats['average'] = req.state.avg
    return {'stats' : stats }
```

Encrypting and decrypting the message body

Another scenario where we need to customize the routing of the endpoints is when we must secure the message body through encryption. The following custom request decrypts an encrypted body using Python's `cryptography` module and the key of the encrypted body:

```
from cryptography.fernet import Fernet


class DecryptRequest(Request):
```

```
async def body(self):
    body = await super().body()
    login_dict = ast.literal_eval(body.decode('utf-8'))
    fernet = Fernet(bytes(login_dict['key'],
        encoding='utf-8'))
    data = fernet.decrypt(
      bytes(login_dict['enc_login'], encoding='utf-8'))
    self.state.dec_data = json.loads(
        data.decode('utf-8'))
    return body
```

> **Important note**
> The `cryptography` module requires the installation of the `itsdangerous` extension for the encryption/decryption procedure used in this project.

`DecryptRequest` will decrypt the message and return the list of login records as a request `state` object. The following service provides the encrypted message body and key and returns the decrypted list of login records from `DecryptRequest` as a response:

```
@router.post("/login/decrypt/details")
async def send_decrypt_login(enc_data: EncLoginReq,
    req:Request, user: str = Depends(get_current_user)):
    return {"data" : req.state.dec_data}
```

Note that `send_decrypt_login()` has an `EncLoginReq` request model that contains the encrypted message body and the encryption key from the client.

Customizing the routes and their `Request` objects can help optimize and streamline microservice transactions, especially those API endpoints that require heavy loads on message body conversions, transformations, and computations.

Now, our next discussion will focus on applying different `Response` types for the API services.

Choosing the appropriate responses

The FastAPI framework offers other options for rendering API endpoint responses other than the most common `JsonResponse` option. Here is a list of some of the response types supported by FastAPI and their corresponding samples from our application:

- The API endpoints can utilize the `PlainTextResponse` type if their response is text-based only. The following `intro_list_restaurants()` service returns a text-based message to the client:

```
@router.get("/restaurant/index")
def intro_list_restaurants():
    return PlainTextResponse(content="The Restaurants")
```

- Services can use `RedirectResponse` if they need to pursue navigation to another entirely different application or another endpoint of the same application. The following endpoint jumps to a hypertext reference about some known Michelin-starred restaurants:

```
@router.get("/restaurant/michelin")
def redirect_restaurants_rates():
    return RedirectResponse(
        url="https://guide.michelin.com/en/restaurants")
```

- A `FileResponse` type can help services render some content of a file, preferably text-based files. The following `load_questions()` service shows the list of questions saved in the `questions.txt` file placed inside the `/file` folder of the application:

```
@router.get("/question/load/questions")
async def load_questions(user: str =
                    Depends(get_current_user)):
    file_path = os.getcwd() +
    '\\files\\questions.txt';
    return FileResponse(path=file_path,
                media_type="text/plain")
```

- `StreamingResponse` is another response type that can provide us with another approach to the **Server-Sent Events** (**SSE**) implementation. *Chapter 8, Creating Coroutines, Events, and Message-Driven Transactions*, has provided us with an SSE that utilizes the `EventSourceResponse` type:

```
@router.get("/question/sse/list")
async def list_questions(req:Request,
        engine=Depends(create_db_engine),
```

```
                    user: str = Depends(get_current_user)):
        async def print_questions():
            repo:QuestionRepository =
                    QuestionRepository(engine)
            result = await repo.get_all_question()
            for q in result:
                disconnected = await req.is_disconnected()
                if disconnected:
                    break
                yield 'data: {}\n\n.format(
                    json.dumps(jsonable_encoder(q),
                        cls=MyJSONEncoder))
                await asyncio.sleep(1)
        return StreamingResponse(print_questions(),
                    media_type="text/event-stream")
```

- Services that render images can also use the StreamingResponse type. The following
  logo_upload_png() service uploads any **JPEG** or **PNG** file and renders it in the browser:

```
@router.post("/restaurant/upload/logo")
async def logo_upload_png(logo: UploadFile = File(...)):
    original_image = Image.open(logo.file)
    original_image =
        original_image.filter(ImageFilter.SHARPEN)

    filtered_image = BytesIO()
    if logo.content_type == "image/png":
        original_image.save(filtered_image, "PNG")
        filtered_image.seek(0)
        return StreamingResponse(filtered_image,
                media_type="image/png")
    elif logo.content_type == "image/jpeg":
        original_image.save(filtered_image, "JPEG")
        filtered_image.seek(0)
        return StreamingResponse(filtered_image,
```

```
                media_type="image/jpeg")
```

- The `StreamingResponse` type is also effective in rendering videos in various formats such as **MP4**. The following service reads a file inside the application named `sample.mp4` and publishes it to the browser:

```
@router.get("/restaurant/upload/video")
def video_presentation():
    file_path = os.getcwd() + '\\files\\sample.mp4'
    def load_file():
        with open(file_path, mode="rb") as video_file:
            yield from video_file
    return StreamingResponse(load_file(),
            media_type="video/mp4")
```

- If the service wants to publish a simple HTML markup page without making references to static CSS or JavaScript files, then `HTMLResponse` is the right choice. The following service renders an HTML page with a Bootstrap framework provided by some CDN libraries:

```
@router.get("/signup")
async def signup(engine=Depends(create_db_engine),
        user: str = Depends(get_current_user) ):
    signup_content = """
    <html lang='en'>
        <head>
            <meta charset="UTF-8">
            <script src="https://code.jquery.com/jquery-
                    3.4.1.min.js"></script>
            <link rel="stylesheet"
              href="https://stackpath.bootstrapcdn.com/
                bootstrap/4.4.1/css/bootstrap.min.css">

            <script src="https://cdn.jsdelivr.net/npm/
                popper.js@1.16.0/dist/umd/popper.min.js">
            </script>
            <script
```

```
                   src="https://stackpath.bootstrapcdn.com/
           bootstrap/4.4.1/js/bootstrap.min.js"></script>

       </head>
       <body>
         <div class="container">
           <h2>Sign Up Form</h2>
           <form>
               <div class="form-group">
                   <label for="firstname">
                          Firstname:</label>
                   <input type='text'
                       class="form-control"
                       name='firstname'
                       id='firstname'/><br/>
               </div>
               ... ... ... ... ... ... ... ...
               <div class="form-group">
                   <label for="role">Role:</label>
                   <input type='text'
                     class="form-control"
                     name='role' id='role'/><br/>
               </div>
               <button type="submit" class="btn
                   btn-primary">Sign Up</button>
           </form>
         </div>
       </body>
</html>
"""

return HTMLResponse(content=signup_content,
          status_code=200)
```

- If the API endpoints have other rendition types needed to be published, the `Response` class can customize them through its `media_type` property. The following is a service that converts JSON data into XML content by setting the `media_type` property of `Response` to the `application/xml` MIME type:

```
@router.get("/keyword/list/all/xml")
async def
    convert_to_xml(engine=Depends(create_db_engine),
        user: str = Depends(get_current_user)):

    repo:KeyRepository = KeyRepository(engine)
    list_of_keywords = await repo.get_all_keyword()
    root = minidom.Document()
    xml = root.createElement('keywords')
    root.appendChild(xml)

    for keyword in list_of_keywords:
        key = root.createElement('keyword')
        word = root.createElement('word')
        key_text = root.createTextNode(keyword.word)
        weight= root.createElement('weight')
        weight_text =
            root.createTextNode(str(keyword.weight))
        word.appendChild(key_text)
        weight.appendChild(weight_text)
        key.appendChild(word)
        key.appendChild(weight)
        xml.appendChild(key)

    xml_str = root.toprettyxml(indent ="\t")
    return Response(content=xml_str,
            media_type="application/xml")
```

Although FastAPI is not a web framework, it can support Jinja2 templating for rare cases where API services require rendering their response as an HTML page. Let us highlight how API services utilize Jinja2 templates as part of the response.

Setting up the Jinja2 template engine

First, we need to install the `jinja2` module using `pip`:

```
pip install jinja2
```

Then, we need to create a folder that will hold all the Jinja2 templates. Jinja2 must define this folder, usually named `templates`, by creating the `Jinja2Templates` instance in `FastAPI` or any `APIRouter`. The following snippet is part of the `/api/login.py` router that shows the setup and configuration of the Jinja2 templating engine:

```
from fastapi.templating import Jinja2Templates


router = APIRouter()
templates = Jinja2Templates(directory="templates")
```

Setting up the static resources

After the `templates` folder, the Jinja2 engine requires the application to have a folder named `static` in the project directory to hold the CSS, JavaScript, images, and other static files for the Jinja2 templates. Then, we need to instantiate the `StaticFiles` instance to define the `static` folder and map it with a virtual name. Additionally, the `StaticFiles` instance must be mounted to a specific path through `FastAPI`'s `mount()` method. We also need to set the `html` property of the `StaticFiles` instance to `True` to set the folder in HTML mode. The following configuration shows how to set up the static resource folder in the `main.py` module:

```
from fastapi.staticfiles import StaticFiles


app.mount("/static", StaticFiles(directory="static",
            html=True), name="static")
```

For the FastAPI components to access these static files, the engine needs the `aiofiles` extension installed:

```
pip install aiofiles
```

Creating the template layout

The following template is the **base** or **parent** template for the application that can now access the Bootstrap resources from the `static` folder due to the template engine and `aiofiles` module:

```
<!DOCTYPE html>
<html lang="en">
    <head>
        <meta charset="UTF-8">
        <meta http-equiv="X-UA-Compatible"
            content="IE=edge">
        <meta name="viewport" content="width=device-width,
            initial-scale=1.0, shrink-to-fit=no">
        <meta name="apple-mobile-web-app-capable"
            content="yes">

        <link rel="stylesheet" type="text/css"
            href="{{url_for('static',
                path='/css/bootstrap.min.css')}}">
        <script src="{{url_for('static', path='/js/
            jquery-3.6.0.js')}}"></script>
        <script src="{{url_for('static',
            path='/js/bootstrap.min.js')}}"></script>
    </head>
    <body>
        {% block content %}
        {% endblock content %}
    </body>
</html>
```

Other templates can inherit the structure and design of this `layout.html` using the `{% extends %}` tags. The Jinja2 base template, like our `layout.html`, has these Jinja2 tags, namely the `{% block content %}` and `{% endblock %}` tags, which indicate where child templates can insert their content during the translation phase. But for all these templates to work, they must be saved in the `/templates` directory. The following is a sample child template named `users.html` that generates a table of profiles from the context data:

```
{% extends "layout.html" %}
```

```
{% block content %}
<div class="container">
<h2>List of users </h2>
<p>This is a Boostrap 4 table applied to JinjaTemplate.</p>
<table class="table">
    <thead>
        <tr>
            <th>Login ID</th>
            <th>Username</th>
            <th>Password</th>
            <th>Passphrase</th>
        </tr>
    </thead>
    <tbody>
    {% for login in data %}
    <tr>
        <td>{{ login.login_id}}</td>
        <td>{{ login.username}}</td>
        <td>{{ login.password}}</td>
        <td>{{ login.passphrase}}</td>
    </tr>
    {% endfor%}
</tbody>
</table>
</div>
{% endblock %}
```

Observe that the child Jinja2 template also has the "block" tags to mark the content to be merged into the parent template.

For the API to render the templates, the service must use the Jinja2 engine's `TemplateResponse` type as the response type. `TemplateResponse` needs the filename of the template, the `Request` object, and the context data if there is any. The following is the API service that renders the previous `users.html` template:

```
@router.get("/login/html/list")
async def list_login_html(req: Request,
        engine=Depends(create_db_engine),
```

```
        user: str = Depends(get_current_user)):
    repo:LoginRepository = LoginRepository(engine)
    result = await repo.get_all_login()
    return templates.TemplateResponse("users.html",
            {"request": req, "data": result})
```

Using ORJSONResponse and UJSONResponse

When it comes to yielding numerous dictionaries or JSON-able-components, it is appropriate to use either ORJSONResponse or UJSONResponse. ORJSONResponse uses orjson to serialize a humongous listing of dictionary objects into a JSON string as a response. So, first, we need to install orjson using the pip command before using ORJSONResponse. ORJSONResponse serializes UUID, numpy, data classes, and datetime objects faster than the common JSONResponse.

However, UJSONResponse is relatively faster than ORJSONResponse because it uses the ujson serializer. The ujson serializer must first be installed before using UJSONResponse.

The following are the two API services that use these two fast alternatives for a JSON serializer:

```
@router.get("/login/list/all")
async def list_all_login(engine=Depends(create_db_engine),
        user: str = Depends(get_current_user)):
    repo:LoginRepository = LoginRepository(engine)
    result = await repo.get_all_login()
    return ORJSONResponse(content=jsonable_encoder(result),
            status_code=201)


@router.get("/login/account")
async def get_login(id:int,
        engine=Depends(create_db_engine),
        user: str = Depends(get_current_user) ):
    repo:LoginRepository = LoginRepository(engine)
    result = await repo.get_login_id(id)
    return UJSONResponse(content=jsonable_encoder(result),
            status_code=201)
```

We still need to apply the jsonable_encoder() component to convert BSON's ObjectId of the result into str before the two responses pursue their serialization processes. Now, let us focus on how we provide internal API documentation using the OpenAPI 3.0 specification.

Applying the OpenAPI 3.x specification

The OpenAPI 3.0 specification is a standard API documentation and language-agnostic specification that can describe the API services without knowing its sources, reading its documentation, and understanding its business logic. Additionally, FastAPI supports OpenAPI, and it can even automatically generate the default internal documentation of the API based on OpenAPI standards.

There are three ways to document our API services using the specification:

- By extending the OpenAPI schema definition
- By using the internal code base properties
- By using the Query, Body, Form, and Path functions

Extending the OpenAPI schema definition

FastAPI has a get_openapi() method from its fastapi.openapi.utils extension that can override some schema descriptions. We can modify the info, servers, and paths details of the schema definition through the get_openapi() function. The function returns a dict of all details of the OpenAPI schema definition of the application.

The default OpenAPI schema documentation is always set up in the main.py module because it is consistently associated with the FastAPI instance. For the function to generate the dict of schema details, it must accept at least the title, version, and routes parameter values. The following custom function extracts the default *OpenAPI* schema for updating:

```python
def update_api_schema():
    DOC_TITLE = "The Online Restaurant Rating System API"
    DOC_VERSION = "1.0"
    openapi_schema = get_openapi(
        title=DOC_TITLE,
        version=DOC_VERSION,
        routes=app.routes,
    )

app.openapi_schema = openapi_schema
return openapi_schema
```

The `title` parameter value is the document title, the `version` parameter value is the version of the API implementation, and `routes` contains a list of registered API services. Observe that the last line before the `return` statement updates FastAPI's built-in `openapi_schema` defaults. Now, to update the general information details, we use the `info` key of the schema definition to change some values, as shown in the following sample:

```
openapi_schema["info"] = {
        "title": DOC_TITLE,
        "version": DOC_VERSION,
        "description": "This application is a prototype.",
        "contact": {
            "name": "Sherwin John Tragura",
            "url": "https://ph.linkedin.com/in/sjct",
            "email": "cowsky@aol.com"
        },
        "license": {
            "name": "Apache 2.0",
            "url": "https://www.apache.org/
                    licenses/LICENSE-2.0.html"
        },
    }
```

The preceding info schema update must also be part of the `update_api_schema()` function together with the update on the documentation of each registered API service. These details can includeAPI service's description and summary, the POST endpoint's description of its `requestBody` and GET endpoint's details about its parameters, and the API tags. Add the following `paths` updates:

```
openapi_schema["paths"]["/ch09/login/authenticate"]["post"]
["description"] = "User Authentication Session"
openapi_schema["paths"]["/ch09/login/authenticate"]["post"]
["summary"] = "This is an API that stores credentials in
session."
openapi_schema["paths"]["/ch09/login/authenticate"]["post"]
["tags"] = ["auth"]

openapi_schema["paths"]["/ch09/login/add"]["post"]
["description"] = "Adding Login User"
openapi_schema["paths"]["/ch09/login/add"]["post"]
```

```
["summary"] = "This is an API adds new user."
openapi_schema["paths"]["/ch09/login/add"]["post"]
["tags"] = ["operation"]
openapi_schema["paths"]["/ch09/login/add"]["post"]
["requestBody"]["description"]="Data for LoginReq"


openapi_schema["paths"]["/ch09/login/profile/add"]
["description"] = "Updating Login User"
openapi_schema["paths"]["/ch09/login/profile/add"]
["post"]["summary"] = "This is an API updating existing user
record."
openapi_schema["paths"]["/ch09/login/profile/add"]
["post"]["tags"] = ["operation"]
openapi_schema["paths"]["/ch09/login/profile/add"]
["post"]["requestBody"]["description"]="Data for LoginReq"


openapi_schema["paths"]["/ch09/login/html/list"]["get"]
["description"] = "Renders Jinja2Template with context data."
openapi_schema["paths"]["/ch09/login/html/list"]["get"]
["summary"] = "Uses Jinja2 template engine for rendition."
openapi_schema["paths"]["/ch09/login/html/list"]["get"]["tags"]
= ["rendition"]


openapi_schema["paths"]["/ch09/login/list/all"]["get"]
["description"] = "List all the login records."
openapi_schema["paths"]["/ch09/login/list/all"]["get"]
["summary"] = "Uses JsonResponse for rendition."
openapi_schema["paths"]["/ch09/login/list/all"]["get"]["tags"]
= ["rendition"]
```

The preceding will give us a new OpenAPI document dashboard, as shown in *Figure 9.1*:

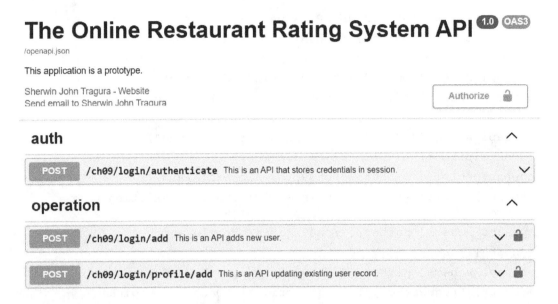

Figure 9.1 – A customized OpenAPI dashboard

Tags are essential variables of the OpenAPI documentation because they organize the API endpoints according to the routers, business processes, requirements, and modules. It is a best practice to use tags.

Once all the updates have been set, replace FastAPI's openapi() function with the new update_api_schema() function.

Using the internal code base properties

FastAPI's constructor has parameters that can replace the default info document details without using the get_openapi() function. The following snippet showcases a sample documentation update on the title, description, version, and servers details of the OpenAPI documentation:

```
app = FastAPI(… … … …,
            title="The Online Restaurant Rating
                    System API",
            description="This a software prototype.",
            version="1.0.0",
            servers= [
                {
                    "url": "http://localhost:8000",
                    "description": "Development Server"
                },
```

```
                  {
                      "url": "https://localhost:8002",
                      "description": "Testing Server",
                  }
              ])
```

When adding documentation to the API endpoints, the path operators of `FastAPI` and `APIRouter` also have parameters that allow changes to the default OpenAPI variables attributed to each endpoint. The following is a sample service that updates its `summary`, `description`, `response_description`, and other response details through the `post()` path operator:

```
@router.post("/restaurant/add",
    summary="This API adds new restaurant details.",
    description="This operation adds new record to the
        database. ",
    response_description="The message body.",
    responses={
        200: {
            "content": {
                "application/json": {
                    "example": {
                        "restaurant_id": 100,
                        "name": "La Playa",
                        "branch": "Manila",
                        "address": "Orosa St.",
                        "province": "NCR",
                        "date_signed": "2022-05-23",
                        "city": "Manila",
                        "country": "Philippines",
                        "zipcode": 1603
                    }
                }
            },
        },
        404: {
            "description": "An error was encountered during
                saving.",
```

```
                    "content": {
                        "application/json": {
                            "example": {"message": "insert login
                                unsuccessful"}
                        }
                    },
                },
            },
        tags=["operation"])
async def add_restaurant(req:RestaurantReq,
        engine=Depends(create_db_engine),
            user: str = Depends(get_current_user)):
    restaurant_dict = req.dict(exclude_unset=True)
    restaurant_json = dumps(restaurant_dict,
            default=json_datetime_serializer)
    repo:RestaurantRepository =
            RestaurantRepository(engine)
    result = await repo.insert_restaurant(
            loads(restaurant_json))
    if result == True:
        return req
    else:
        return JSONResponse(content={"message":
            "insert login unsuccessful"}, status_code=500)
```

Using the Query, Form, Body, and Path functions

Aside from the declaration and additional validations, the Query, Path, Form, and Body parameter functions can also be used to add some metadata to the API endpoints. The following authenticate() endpoint has added descriptions and validations through the Query() function:

```
@router.post("/login/authenticate")
async def authenticate(response: Response,
    username:str = Query(...,
        description='The username of the credentials.',
        max_length=50),
    password: str = Query(...,
```

```
    description='The password of the of the credentials.',
    max_length=20),
  engine=Depends(create_db_engine)):
  repo:LoginRepository = LoginRepository(engine)
  ... ... ... ... ... ...
  response.set_cookie("session", token)
  return {"username": username}
```

The following get_login() uses the Path() directive to insert a description of the id parameter:

```
@router.get("/login/account/{id}")
async def get_login(id:int = Path(...,
         description="The user ID of the user."),
   engine=Depends(create_db_engine),
   user: str = Depends(get_current_user) ):
   ... ... ... ... ... ...
   return UJSONResponse(content=jsonable_encoder(result),
         status_code=201)
```

The description and max_length metadata of the Query() function will become part of the OpenAPI documentation for authenticate(), as shown in *Figure 9.2*:

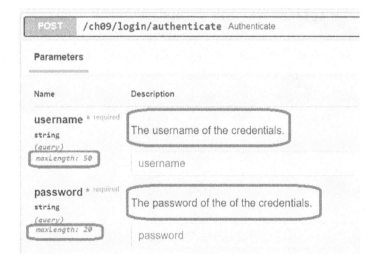

Figure 9.2 – The Query metadata

Additionally, the description metadata of the Path() directive will also appear in the get_login() documentation, as presented in *Figure 9.3*:

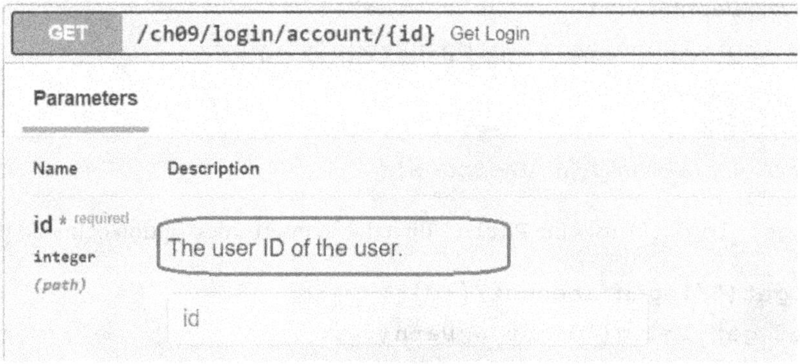

Figure 9.3 – The Path metadata

Likewise, we can add descriptions to form parameters using the Form directive. The following service shows you how to insert documentation through the Form directive:

```python
@router.post("/user/profile")
async def create_profile(req: Request,
        firstname: str = Form(...,
          description='The first name of the user.'),
        lastname: str = Form(...,
          description='The last name of the user.'),
        age: int = Form(...,
          description='The age of the user.'),
        birthday: date = Form(...,
           description='The birthday of the user.'),
        user: str = Depends(get_current_user)):
    user_details = req.session["user_details"]
    return {'profile' : user_details}
```

Moreover, it is also possible to document all types of HTTP responses or status codes that the API service can throw through the path operator's `responses` parameter. The following `video_presentation()` service provides metadata regarding the nature of its response when it encounters no errors (*HTTP Status Code 200*) and with runtime errors (*HTTP Status Code 500*):

```
from models.documentation.response import Error500Model
... ... ... ... ...
@router.get("/restaurant/upload/video",responses={
        200: {
            "content": {"video/mp4": {}},
            "description": "Return an MP4 encoded video.",
        },
        500:{
            "model": Error500Model,
            "description": "The item was not found"
        }
    },)
def video_presentation():
    file_path = os.getcwd() + '\\files\\sample.mp4'
    def load_file():
        with open(file_path, mode="rb") as video_file:
            yield from video_file
    return StreamingResponse(load_file(),
            media_type="video/mp4")
```

`Error500Model` is a `BaseModel` class that will give you a clear picture of the response once the application encounters an *HTTP Status Code 500* error and will only be used in the OpenAPI documentation. It contains metadata such as the message that holds a hardcoded error message. *Figure 9.4* shows the resulting OpenAPI documentation for `video_presentation()` after adding the metadata for its responses:

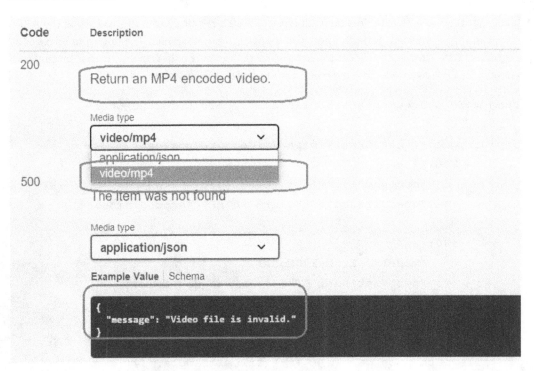

Code	Description
200	Return an MP4 encoded video.
	Media type
	video/mp4 ⌄
	application/json
	video/mp4
500	The item was not found
	Media type
	application/json ⌄
	Example Value Schema
	{ "message": "Video file is invalid." }

Figure 9.4 – The documentation for API responses

And now, for our last discussion, let us explore how we can perform unit testing in FastAPI, which could lead to a test-driven development setup.

Testing the API endpoints

FastAPI uses the pytest framework to run its test classes. So, before we create our test classes, first, we need to install the pytest framework using the pip command:

```
pip install pytest
```

FastAPI has a module called fastapi.testclient where all components are Request-based, including the TestClient class. To access all the API endpoints, we need the TestClient object. But first, we need to create a folder such as test, which will contain test modules where we implement our test methods. We place our test methods outside main.py or the router modules to maintain clean code and organization.

Writing the unit test cases

It is a best practice to write one test module per router component, except for cases where there is a tight connection between these routers. We place these test modules inside the `test` directory. To pursue the automated testing, we need to import the `APIRouter` instance or the `FastAPI` instance into the test module to set up `TestClient`. `TestClient` is almost like Python's client module, `requests`, when it comes to the helper methods used to consume APIs.

The method names of the test cases must start with a `test_` prefix, which is a `pytest` requirement. Test methods are all standard Python methods and should not be asynchronous. The following is a test method in `test/test_restaurants.py` that checks whether the endpoint returns the proper text-based response:

```
from fastapi.testclient import TestClient


from api import restaurant


client = TestClient(restaurant.router)


def test_restaurant_index():
    response = client.get("/restaurant/index")
    assert response.status_code == 200
    assert response.text == "The Restaurants"
```

`TestClient` supports assert statements that check the response of its helper methods, like `get()`, `post()`, `put()`, and `delete()` the status code and response body of the API. The `test_restaurant_index()`, for instance, uses the `get()` method of the TestClient API to run `/restaurant/index` GET service and extract its response. The assert statements are used if the `statuc_code` and `response.text` are correct. The endpoint has no imposed dependencies, so the test module is *router-based*.

Mocking the dependencies

Testing API endpoints with dependencies is not as straightforward as the previous example. Our endpoints have session-based security through the JWT and the `APIKeyCookie` class, so we cannot just run `pytest` to test them. First, we need to apply *mocking* to these dependencies by adding them to the `dependency_overrides` of the `FastAPI` instance. Since `APIRouter` cannot mock dependencies, we need to use the `FastAPI` instance to set up `TestClient`. All endpoints can be unit tested if the routers are part of the FastAPI configuration through `include_router()`:

```
from fastapi.testclient import TestClient
from models.data.orrs import Login
```

```
from main import app

from util.auth_session import get_current_user

client = TestClient(app)

async def get_user():
    return Login(**{"username": "sjctrags",
        "login_id": 101,
        "password":"sjctrags", "passphrase": None,
        "profile": None})

app.dependency_overrides[get_current_user] =  get_user

def test_rating_top_three():
    response = client.post("/ch09/rating/top/three",
    json={
            "rate1": 10.0,
            "rate2": 20.0 ,
            "rate3": 30.0

    })
    assert response.status_code == 200
    assert response.json() == { "stats": {
            "sum": 60.0,
            "average": 20.0

    }
}
```

The /rating/top/three API from the /api/route_extract.py router requires a dict of ratings to derive a JSON result containing average and sum. TestClient's path operators have JSON and data parameters, where we can pass test data to the API. Likewise, TestClient's response has methods that can derive the expected response body, such as, in this example, the json() function.

Running the test method will result in some `APIKeyCookie` exceptions due to the dependency on session-based security. To bypass this issue, we need to create a fake `get_current_user()` dependable function to proceed with the testing. We add the `get_current_user()` dependable function into the roster of overrides and map it with the fake ones, such as our `get_user()` function, to replace its execution. This process is what we call **mocking** in the FastAPI context.

Aside from security, we can also mock the database connection by creating a mock database object or database engine, depending on whether it is a relational database or a NoSQL database. In the following test case, we are performing a unit test in `/ch09/login/list/all`, which needs MongoDB connectivity to access the list of login profiles. For the test to work, we need to create a mock `AsyncIOMotorClient` object with a dummy test database called `orrs_test`. Here is the `test_list_login()` test, which implements this database mocking:

```
def db_connect():
    client_od =
        AsyncIOMotorClient(f"mongodb://localhost:27017/")
    engine = AIOEngine(motor_client=client_od,
            database="orrs_test")
    return engine


async def get_user():
    return Login(**{"username": "sjctrags", "login_id": 101,
            "password":"sjctrags", "passphrase": None,
            "profile": None})


app.dependency_overrides[get_current_user] =  get_user
app.dependency_overrides[create_db_engine] = db_connect


def test_list_login():
    response = client.get("/ch09/login/list/all")
    assert response.status_code == 201
```

Running test methods

Run the `pytest` command on the command line to execute all unit tests. The `pytest` engine will compile and run all `TestClient` apps in the `test` folder, thus running all the test methods. *Figure 9.5* shows a snapshot of the test result:

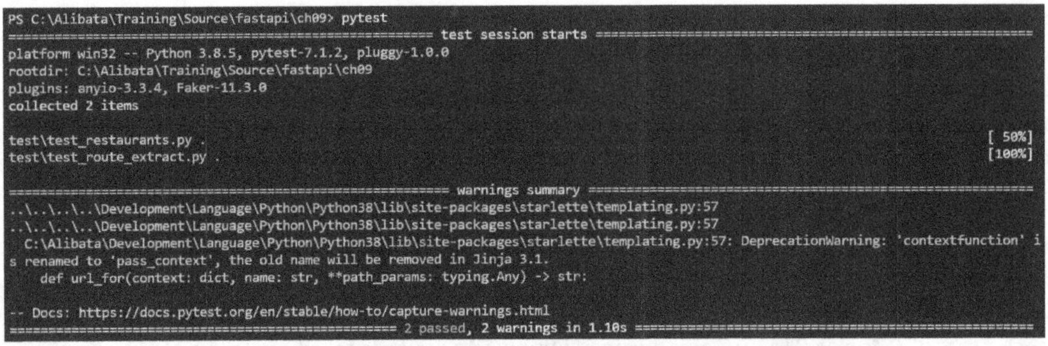

```
PS C:\Alibata\Training\Source\fastapi\ch09> pytest
========================================= test session starts =========================================
platform win32 -- Python 3.8.5, pytest-7.1.2, pluggy-1.0.0
rootdir: C:\Alibata\Training\Source\fastapi\ch09
plugins: anyio-3.3.4, Faker-11.3.0
collected 2 items

test\test_restaurants.py .                                                                      [ 50%]
test\test_route_extract.py .                                                                    [100%]

========================================== warnings summary ===========================================
..\..\..\..\Development\Language\Python\Python38\lib\site-packages\starlette\templating.py:57
..\..\..\..\Development\Language\Python\Python38\lib\site-packages\starlette\templating.py:57
  C:\Alibata\Development\Language\Python\Python38\lib\site-packages\starlette\templating.py:57: DeprecationWarning: 'contextfunction' i
s renamed to 'pass_context', the old name will be removed in Jinja 3.1.
    def url_for(context: dict, name: str, **path_params: typing.Any) -> str:

-- Docs: https://docs.pytest.org/en/stable/how-to/capture-warnings.html
=========================================== 2 passed, 2 warnings in 1.10s =============================
```

Figure 9.5 – The test result

Learning more about the `pytest` framework offers a heads-up in understanding the automation of test cases in FastAPI. Organizing all test methods through modules is essential in the testing phase of the application since we run all of them in bulk.

Summary

This chapter showcased some essential features that were not part of the previous chapters but can help fill some gaps during microservice development. One involves choosing better and more appropriate JSON serializers and de-serializers when converting a huge amount of data into JSON. Also, the advanced customizations, session handling, message body encryption and decryption, and testing API endpoints gave us a clear understanding of the potential of FastAPI to create cutting-edge and progressive microservice solutions. Also, this chapter introduced different API responses supported by FastAPI, including Jinja2's `TemplateResponse`.

The next chapter will show us the strength of FastAPI in cracking numerical and symbolic computations.

10

Solving Numerical, Symbolic, and Graphical Problems

Microservice architecture is not only used to build fine-grained, optimized, and scalable applications in the banking, insurance, production, human resources, and manufacturing industries. It is also used to develop scientific and computation-related research and scientific software prototypes for applications such as **laboratory information management systems** (**LIMSs**), weather forecasting systems, **geographical information systems** (**GISs**), and healthcare systems.

FastAPI is one of the best choices in building these granular services since they usually involve highly computational tasks, workflows, and reports. This chapter will highlight some transactions not yet covered in the previous chapters, such as symbolic computations using `sympy`, solving linear systems using `numpy`, plotting mathematical models using `matplotlib`, and generating data archives using `pandas`. This chapter will also show you how FastAPI is flexible when solving workflow-related transactions by simulating some Business Process Modeling Notation (BPMN) tasks. For developing big data applications, a portion of this chapter will showcase GraphQL queries for big data applications and Neo4j graph databases for graph-related projects with the framework.

The main objective of this chapter is to introduce the FastAPI framework as a tool for providing microservice solutions for scientific research and computational sciences.

In this chapter, we will cover the following topics:

- Setting up the projects
- Implementing the symbolic computations
- Creating arrays and DataFrames
- Performing statistical analysis
- Generating CSV and XLSX reports
- Plotting data models

- Simulating a BPMN workflow
- Using GraphQL queries and mutations
- Utilizing the Neo4j graph database

Technical requirements

This chapter provides the base skeleton of a **periodic census and computational system** that enhances fast data collection procedures in different areas of a specific country. Although unfinished, the prototype provides FastAPI implementations that highlight important topics of this chapter, such as creating and plotting mathematical models, gathering answers from respondents, providing questionnaires, creating workflow templates, and utilizing a graph database. The code for this chapter can be found at https://github.com/PacktPublishing/Building-Python-Microservices-with-FastAPI in the ch10 project.

Setting up the projects

The PCCS project has two versions: ch10-relational, which uses a PostgreSQL database with Piccolo ORM as the data mapper, and ch10-mongo, which saves data as MongoDB documents using Beanie ODM.

Using the Piccolo ORM

ch10-relational uses a fast Piccolo ORM that can support both sync and async CRUD transactions. This ORM was not introduced in *Chapter 5, Connecting to a Relational Database*, because it is more appropriate for computational, data science-related, and big data applications. The Piccolo ORM is different from other ORMs because it scaffolds a project containing the initial project structure and templates for customization. But before creating the project, we need to install the piccolo module using pip:

```
pip install piccolo
```

Afterward, install the piccolo-admin module, which provides helper classes for the GUI administrator page of its projects:

```
pip install piccolo-admin
```

Now, we can create a project inside a newly created root project folder by running piccolo asgi new, a CLI command that scaffolds the Piccolo project directory. The process will ask for the API framework and application server to utilize, as shown in the following screenshot:

```
Command Prompt                                                                    _  □  x
C:\Alibata\Training\Source\fastapi\ch10>piccolo asgi new
Can't import the APP_REGISTRY from piccolo_conf - some commands may be missing. If this is a new proje
ct don't worry. To see a full traceback use `piccolo --diagnose`
Which routing framework?
starlette [0], fastapi [1], blacksheep [2], xpresso [3]
1
Which server?
uvicorn [0], Hypercorn [1]
0
Run `pip install -r requirements.txt` and `python main.py` to get started.
```

Figure 10.1 – Scaffolding a Piccolo ORM project

You must use FastAPI for the application framework and `uvicorn` is the recommended ASGI server. Now, we can add Piccolo applications inside the project by running the `piccolo app new` command inside the project folder. The following screenshot shows the main project directory, where we execute the CLI command to create a Piccolo application:

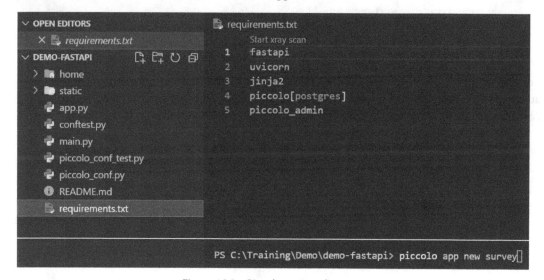

Figure 10.2 – Piccolo project directory

The scaffolded project always has a default application called home, but it can be modified or even deleted. Once removed, the Piccolo platform allows you to replace home by adding a new application to the project by running the `piccolo app new` command inside the project folder, as shown in the preceding screenshot. A Piccolo application contains the ORM models, BaseModel, services, repository classes, and API methods. Each application has an auto-generated `piccolo_app.py` module where we need to configure an `APP_CONFIG` variable to register all the ORM details. The following is the configuration of our project's survey application:

```
APP_CONFIG = AppConfig(
    app_name="survey",
    migrations_folder_path=os.path.join(
        CURRENT_DIRECTORY, "piccolo_migrations"
    ),
    table_classes=[Answers, Education, Question, Choices,
        Profile, Login, Location, Occupation, Respondent],
    migration_dependencies=[],
    commands=[],
)
```

For the ORM platform to recognize the new Piccolo application, its `piccolo_app.py` must be added to `APP_REGISTRY` of the main project's `piccolo_conf.py` module. The following is the content of the `piccolo_conf.py` file of our `ch10-piccolo` project:

```
from piccolo.engine.postgres import PostgresEngine
from piccolo.conf.apps import AppRegistry

DB = PostgresEngine(
    config={
        "database": "pccs",
        "user": "postgres",
        "password": "admin2255",
        "host": "localhost",
        "port": 5433,
    }
)
```

```
APP_REGISTRY = AppRegistry(
    apps=["survey.piccolo_app",
        "piccolo_admin.piccolo_app"]
)
```

The `piccolo_conf.py` file is also the module where we establish the PostgreSQL database connection. Aside from PostgreSQL, the Piccolo ORM also supports SQLite databases.

Creating the data models

Like in Django ORM, Piccolo ORM has migration commands to generate the database tables based on model classes. But first, we need to create model classes by utilizing its `Table` API class. It also has helper classes to establish column mappings and foreign key relationships. The following are some data model classes that comprise our database `pccs`:

```
from piccolo.columns import ForeignKey, Integer, Varchar,
        Text, Date, Boolean, Float
from piccolo.table import Table

class Login(Table):
    username = Varchar(unique=True)
    password = Varchar()

class Education(Table):
    name = Varchar()

class Profile(Table):
    fname = Varchar()
    lname = Varchar()
    age = Integer()
    position = Varchar()
    login_id = ForeignKey(Login, unique=True)
    official_id = Integer()
    date_employed = Date()
```

After creating the model classes, we can update the database by creating the migrations files. Migration is a way of updating the database of a project. In the Piccolo platform, we can run the `piccolo migrations new <app_name>` command to generate files in the `piccolo_migrations` folder. These are called migration files and they contain migration scripts. But to save time, we will include the `--auto` option for the command to let the ORM check the recently executed migration files and auto-generate the migration script containing the newly reflected schema updates. Check the newly created migration file first before running the `piccolo migrations forward <app_name>` command to execute the migration script. This last command will auto-create all the tables in the database based on the model classes.

Implementing the repository layer

Creating the repository layer comes after performing all the necessary migrations. Piccolo's CRUD operations are like those in the Peewee ORM. It is swift, short, and easy to implement. The following code shows an implementation of the `insert_respondent()` transaction, which adds a new respondent profile:

```python
from survey.tables import Respondent
from typing import Dict, List, Any


class RespondentRepository:
    async def insert_respondent(self,
            details:Dict[str, Any]) -> bool:
        try:
            respondent = Respondent(**details)
            await respondent.save()
        except Exception as e:
            return False
        return True
```

Like Peewee, Piccolo's model classes can persist records, as shown by `insert_respondent()`, which implements an asynchronous INSERT transaction. On the other hand, `get_all_respondent()` retrieves all respondent profiles and has the same approach as Peewee, as shown here:

```python
    async def get_all_respondent(self):
        return await Respondent.select()
                    .order_by(Respondent.id)
```

The remaining Peewee-like DELETE and UPDATE respondent transactions are created in the project's `/survey/repository/respondent.py` module.

The Beanie ODM

The second version of the PCCS project, `ch10-mongo`, utilizes a MongoDB datastore and uses the Beanie ODM to implement its asynchronous CRUD transactions. We covered Beanie in *Chapter 6, Using a Non-Relational Database*. Now, let us learn how to apply FastAPI in symbolic computations. We will be using the `ch10-piccolo` project for this.

Implementing symbolic computations

Symbolic computation is a mathematical approach to solving problems using symbols or mathematical variables. It uses mathematical equations or expressions formulated using symbolic variables to solve linear and nonlinear systems, rational expressions, logarithmic expressions, and other complex real-world models. To perform symbolic computation in Python, you must install the `sympy` module using the `pip` command:

```
pip install sympy
```

Let us now start creating our first symbolic expressions.

Creating symbolic expressions

One way of implementing the FastAPI endpoint that performs symbolic computation is to create a service that accepts a mathematical model or equation as a string and converts that string into a `sympy` symbolic expression. The following `substitute_eqn()` processes an equation in `str` format and converts it into valid linear or nonlinear bivariate equations with the `x` and `y` variables. It also accepts values for `x` and `y` to derive the solution of the expression:

```python
from sympy import symbols, sympify


@router.post("/sym/equation")
async def substitute_bivar_eqn(eqn: str, xval:int,
                yval:int):
    try:
        x, y = symbols('x, y')
        expr = sympify(eqn)
        return str(expr.subs({x: xval, y: yval}))
    except:
        return JSONResponse(content={"message":
            "invalid equations"}, status_code=500)
```

Before converting a string equation into a sympy expression, we need to define the x and y variables as Symbols objects using the symbols() utility. This method accepts a string of comma-delimited variable names and returns a tuple of symbols equivalent to the variables. After creating all the needed Symbols() objects, we can convert our equation into sympy expressions by using any of the following sympy methods:

- sympify(): This uses eval() to convert the string equation into a valid sympy expression with all Python types converted into their sympy equivalents

- parse_expr(): A full-fledged expression parser that transforms and modifies the tokens of the expression and converts them into their sympy equivalents

Since the substitute_bivar_eqn() service utilizes the sympify() method, the string expression needs to be sanitized from unwanted code before sympifying to avoid any compromise.

On the other hand, the sympy expression object has a subs() method to substitute values to derive the solution. Its resulting object must be converted into str format for Response to render the data. Otherwise, Response will raise ValueError, regarding the result as non-iterable.

Solving linear expressions

The sympy module allows you to implement services that solve multivariate systems of linear equations. The following API service highlights an implementation that accepts two bivariate linear models in string format with their respective solutions:

```
from sympy import Eq, symbols, Poly, solve, sympify


@router.get("/sym/linear")
async def solve_linear_bivar_eqns(eqn1:str,
            sol1: int, eqn2:str, sol2: int):
    x, y = symbols('x, y')

    expr1 = parse_expr(eqn1, locals())
    expr2 = parse_expr(eqn2, locals())

    if Poly(expr1, x).is_linear and
                Poly(expr1, x).is_linear:
        eq1 = Eq(expr1, sol1)
        eq2 = Eq(expr2, sol2)
        sol = solve([eq1, eq2], [x, y])
        return str(sol)
```

```
        else:
            return None
```

The `solve_linear_bivar_eqns()` service accepts two bivariate linear equations and their respective outputs (or intercepts) and aims to establish a system of linear equations. First, it registers the x and y variables as `sympy` objects and then uses the `parser_expr()` method to transform the string expressions into their `sympy` equivalents. Afterward, the service needs to establish linear equality of these equations using the `Eq()` solver, which maps each `sympy` expression to its solution. Then, the API service passes all these linear equations to the `solve()` method to derive the x and y values. The result of `solve()` also needs to be rendered as a string, like in the substitution.

Aside from the `solve()` method, the API also uses the `Poly()` utility to create a polynomial object from an expression to be able to access essential properties of an equation, such as `is_linear()`.

Solving non-linear expressions

The previous `solve_linear_bivar_eqns()` can be reused to solve non-linear systems. The tweak is to shift the validation from filtering the linear equations to any non-linear equations. The following script highlights this code change:

```
@router.get("/sym/nonlinear")
async def solve_nonlinear_bivar_eqns(eqn1:str, sol1: int,
            eqn2:str, sol2: int):
    … … … … … …

    … … … … … …

    if not Poly(expr1, x, y).is_linear or
            not Poly(expr1, x, y).is_linear:
    … … … … … …

    … … … … … …

        return str(sol)
    else:
        return None
```

Solving linear and non-linear inequalities

The `sympy` module supports solving solutions for both linear and non-linear inequalities but on univariate equations only. The following is an API service that accepts a univariate string expression with its output or intercepts, and extracts the solution using the `solve()` method:

```
@router.get("/sym/inequality")
async def solve_univar_inequality(eqn:str, sol:int):
```

```
x= symbols('x')
expr1 = Ge(parse_expr(eqn, locals()), sol)
sol = solve([expr1], [x])
return str(sol)
```

The sympy module has Gt() or StrictGreaterThan, Lt() or StrictLessThan, Ge() or GreaterThan, and Le() or LessThan solvers, which we can use to create inequality. But first, we need to convert the str expression into a Symbols() object using the parser_expr() method before passing them to these solvers. The preceding service uses the GreaterThan solver, which creates an equation where the left-hand side of the expression is generally larger than the left.

Most applications designed and developed for mathematical modeling and data science use sympy to create complex mathematical models symbolically, plot data directly from the sympy equation, or generate results based on datasets or live data. Now, let us proceed to the next group of API services, which deals with data analysis and manipulation using numpy, scipy, and pandas.

Creating arrays and DataFrames

When numerical algorithms require some arrays to store data, a module called **NumPy**, short for **Numerical Python**, is a good resource for utility functions, objects, and classes that are used to create, transform, and manipulate arrays.

The module is best known for its n-dimensional arrays or ndarrays, which consume less memory storage than the typical Python lists. An ndarray incurs less overhead when performing data manipulation than executing the list operations in totality. Moreover, ndarray is strictly heterogeneous, unlike Python's list collections.

But before we start our NumPy-FastAPI service implementation, we need to install the numpy module using the pip command:

```
pip install numpy
```

Our first API service will process some survey data and return it in ndarray form. The following get_respondent_answers() API retrieves a list of survey data from PostgreSQL through Piccolo and transforms the list of data into an ndarray:

```
from survey.repository.answers import AnswerRepository
from survey.repository.location import LocationRepository
import ujson
import numpy as np


@router.get("/answer/respondent")
async def get_respondent_answers(qid:int):
    repo_loc = LocationRepository()
```

```
    repo_answers = AnswerRepository()
locations = await repo_loc.get_all_location()
data = []
for loc in locations:
    loc_q = await repo_answers
        .get_answers_per_q(loc["id"], qid)
    if not len(loc_q) == 0:
        loc_data = [ weights[qid-1]
            [str(item["answer_choice"])]
            for item in loc_q]
        data.append(loc_data)
    arr = np.array(data)
    return ujson.loads(ujson.dumps(arr.tolist()))
```

Depending on the size of the data retrieved, it would be faster if we apply the `ujson` or `orjson` serializers and de-serializers to convert `ndarray` into JSON data. Even though numpy has data types such as `uint`, `single`, `double`, `short`, `byte`, and `long`, JSON serializers can still manage to convert them into their standard Python equivalents. Our given API sample prefers `ujson` utilities to convert the array into a JSON-able response.

Aside from NumPy, `pandas` is another popular module that's used in data analysis, manipulation, transformation, and retrieval. But to use pandas, we need to install NumPy, followed by the `pandas`, `matplotlib`, and `openpxyl` modules:

```
pip install pandas matplotlib openpxyl
```

Let us now discuss about the ndarray in numpy module.

Applying NumPy's linear system operations

Data manipulation in an `ndarray` is easier and faster, unlike in a list collection, which requires list comprehension and loops. The vectors and matrices created by numpy have operations to manipulate their items, such as scalar multiplication, matrix multiplication, transposition, vectorization, and reshaping. The following API service shows how the product between a scalar gradient and an array of survey data is derived using the numpy module:

```
@router.get("/answer/increase/{gradient}")
async def answers_weight_multiply(gradient:int, qid:int):
    repo_loc = LocationRepository()
    repo_answers = AnswerRepository()
    locations = await repo_loc.get_all_location()
    data = []
```

```
    for loc in locations:
        loc_q = await repo_answers
            .get_answers_per_q(loc["id"], qid)
        if not len(loc_q) == 0:
            loc_data = [ weights[qid-1]
              [str(item["answer_choice"])]
                 for item in loc_q]
            data.append(loc_data)
    arr = np.array(list(itertools.chain(*data)))
    arr = arr * gradient
    return ujson.loads(ujson.dumps(arr.tolist()))
```

As shown in the previous scripts, all `ndarray` instances resulting from any numpy operations can be serialized as JSON-able components using various JSON serializers. There are other linear algebraic operations that numpy can implement without sacrificing the performance of the microservice application. Let us take a look now on panda's DataFrame.

Applying the pandas module

In this module, datasets are created as a `DataFrame` object, similar to in Julia and R. It contains rows and columns of data. FastAPI can render these DataFrames using any JSON serializers. The following API service retrieves all survey results from all survey locations and creates a DataFrame from these datasets:

```
import ujson
import numpy as np
import pandas as pd


@router.get("/answer/all")
async def get_all_answers():
    repo_loc = LocationRepository()
    repo_answers = AnswerRepository()
    locations = await repo_loc.get_all_location()
    temp = []
    data = []
    for loc in locations:
        for qid in range(1, 13):
            loc_q1 = await repo_answers
```

```
                    .get_answers_per_q(loc["id"], qid)
            if not len(loc_q1) == 0:
                loc_data = [ weights[qid-1]
                    [str(item["answer_choice"])]
                        for item in loc_q1]
                temp.append(loc_data)
        temp = list(itertools.chain(*temp))
        if not len(temp) == 0:
            data.append(temp)
        temp = list()
    arr = np.array(data)
    return ujson.loads(pd.DataFrame(arr)
        .to_json(orient='split'))
```

The `DataFrame` object has a `to_json()` utility method, which returns a JSON object with an option to format the resulting JSON according to the desired type. On another note, `pandas` can also generate time series, a one-dimensional array depicting a column of a DataFrame. Both DataFrames and time series have built-in methods that are useful for adding, removing, updating, and saving the datasets to CSV and XLSX files. But before we discuss pandas' data transformation processes, let us look at another module that works with `numpy` in many statistical computations, differentiation, integration, and linear optimizations: the `scipy` module.

Performing statistical analysis

The `scipy` module uses numpy as its base module, which is why installing `scipy` requires numpy to be installed first. We can use the `pip` command to install the module:

```
pip install scipy
```

Our application uses the module to derive the declarative statistics of the survey data. The following `get_respondent_answers_stats()` API service computes the mean, variance, skewness, and kurtosis of the dataset using the `describe()` method from `scipy`:

```
from scipy import stats


def ConvertPythonInt(o):
    if isinstance(o, np.int32): return int(o)
    raise TypeError


@router.get("/answer/stats")
```

```
async def get_respondent_answers_stats(qid:int):
    repo_loc = LocationRepository()
    repo_answers = AnswerRepository()
    locations = await repo_loc.get_all_location()
    data = []
    for loc in locations:
        loc_q = await repo_answers
            .get_answers_per_q(loc["id"], qid)
            if not len(loc_q) == 0:
                loc_data = [ weights[qid-1]
                    [str(item["answer_choice"])]
                        for item in loc_q]
            data.append(loc_data)
    result = stats.describe(list(itertools.chain(*data)))
    return json.dumps(result._asdict(),
                default=ConvertPythonInt)
```

The describe() method returns a DescribeResult object, which contains all the computed results. To render all the statistics as part of Response, we can invoke the as_dict() method of the DescribeResult object and serialize it using the JSON serializer.

Our API sample also uses additional utilities such as the chain() method from itertools to flatten the list of data and a custom converter, ConvertPythonInt, to convert NumPy's int32 types into Python int types. Now, let us explore how to save data to CSV and XLSX files using the pandas module.

Generating CSV and XLSX reports

The DataFrame object has built-in to_csv() and to_excel() methods that save its data in CSV or XLSX files, respectively. But the main goal is to create an API service that will return these files as responses. The following implementation shows how a FastAPI service can return a CSV file containing a list of respondent profiles:

```
from fastapi.responses import StreamingResponse
import pandas as pd
from io import StringIO
from survey.repository.respondent import
        RespondentRepository
```

```python
@router.get("/respondents/csv", response_description='csv')
async def create_respondent_report_csv():
    repo = RespondentRepository()
    result = await repo.get_all_respondent()

    ids = [ item["id"] for item in result ]
    fnames = [ f'{item["fname"]}' for item in result ]
    lnames = [ f'{item["lname"]}' for item in result ]
    ages = [ item["age"] for item in result ]
    genders = [ f'{item["gender"]}' for item in result ]
    maritals = [ f'{item["marital"]}' for item in result ]

    dict = {'Id': ids, 'First Name': fnames,
            'Last Name': lnames, 'Age': ages,
            'Gender': genders, 'Married?': maritals}

    df = pd.DataFrame(dict)
    outFileAsStr = StringIO()
    df.to_csv(outFileAsStr, index = False)
    return StreamingResponse(
        iter([outFileAsStr.getvalue()]),
        media_type='text/csv',
        headers={
            'Content-Disposition':
              'attachment;filename=list_respondents.csv',
            'Access-Control-Expose-Headers':
                'Content-Disposition'
        }
    )
```

We need to create a dict() containing columns of data from the repository to create a DataFrame object. From the given script, we store each data column in a separate list(), add all the lists in dict() with keys as column header names, and pass dict() as a parameter to the constructor of DataFrame.

After creating the `DataFrame` object, invoke the `to_csv()` method to convert its columnar dataset into a text stream, `io.StringIO`, which supports Unicode characters. Finally, we must render the `StringIO` object through FastAPI's `StreamResponse` with the `Content-Disposition` header set to rename the default filename of the CSV object.

Instead of using the pandas `ExcelWriter`, our Online Survey application opted for another way of saving `DataFrame` through the `xlsxwriter` module. This module has a `Workbook` class, which creates a workbook containing worksheets where we can plot all column data per row. The following API service uses this module to render XLSX content:

```python
import xlsxwriter
from io import BytesIO


@router.get("/respondents/xlsx",
          response_description='xlsx')
async def create_respondent_report_xlsx():
    repo = RespondentRepository()
    result = await repo.get_all_respondent()
    output = BytesIO()
    workbook = xlsxwriter.Workbook(output)
    worksheet = workbook.add_worksheet()
    worksheet.write(0, 0, 'ID')
    worksheet.write(0, 1, 'First Name')
    worksheet.write(0, 2, 'Last Name')
    worksheet.write(0, 3, 'Age')
    worksheet.write(0, 4, 'Gender')
    worksheet.write(0, 5, 'Married?')
    row = 1
    for respondent in result:
        worksheet.write(row, 0, respondent["id"])

        ... ... ... ... ... ...
        worksheet.write(row, 5, respondent["marital"])
        row += 1
    workbook.close()
    output.seek(0)

    headers = {
        'Content-Disposition': 'attachment;
```

```
            filename="list_respondents.xlsx"'
    }
    return StreamingResponse(output, headers=headers)
```

The given `create_respondent_report_xlsx()` service retrieves all the respondent records from the database and plots each profile record per row in the worksheet from the newly created `Workbook`. Instead of writing to a file, `Workbook` will store its content in a byte stream, `io.ByteIO`, which will be rendered by `StreamResponse`.

The `pandas` module can also help FastAPI services read CSV and XLSX files for rendition or data analysis. It has a `read_csv()` that reads data from a CSV file and converts it into JSON content. The `io.StringIO` stream object will contain the full content, including its Unicode characters. The following service retrieves the content of a valid CSV file and returns JSON data:

```
@router.post("/upload/csv")
async def upload_csv(file: UploadFile = File(...)):
    df = pd.read_csv(StringIO(str(file.file.read(),
            'utf-8')), encoding='utf-16')
    return orjson.loads(df.to_json(orient='split'))
```

There are two ways to handle `multipart` file uploads in FastAPI:

- Use `bytes` to contain the file
- Use `UploadFile` to wrap the file object

Chapter 9, Utilizing Other Advanced Features, introduced the `UploadFile` class for capturing uploaded files because it supports more Pydantic features and has built-in operations that can work with coroutines. It can handle large file uploads without raising an change to - exception when the uploading process reaches the memory limit, unlike using the `bytes` type for file content storage. Thus, the given `read-csv()` service uses `UploadFile` to capture any CSV files for data analysis with `orjson` as its JSON serializer.

Another way to handle file upload transactions is through Jinja2 form templates. We can use `TemplateResponse` to pursue file uploading and render the file content using the Jinja2 templating language. The following service reads a CSV file using `read_csv()` and serializes it into HTML table-formatted content:

```
@router.get("/upload/survey/form",
            response_class = HTMLResponse)
def upload_survey_form(request:Request):
    return templates.TemplateResponse("upload_survey.html",
            {"request": request})
```

```
@router.post("/upload/survey/form")
async def submit_survey_form(request: Request,
          file: UploadFile = File(...)):
    df = pd.read_csv(StringIO(str(file.file.read(),
              'utf-8')), encoding='utf-8')
    return templates.TemplateResponse('render_survey.html',
        {'request': request, 'data': df.to_html()})
```

Aside from `to_json()` and `to_html()`, the `TextFileReader` object also has other converters that can help FastAPI render various content types, including `to_latex()`, `to_excel()`, `to_hdf()`, `to_dict()`, `to_pickle()`, and `to_xarray()`. Moreover, the `pandas` module has a `read_excel()` that can read XLSX content and convert it into any rendition type, just like its `read_csv()` counterpart.

Now, let us explore how FastAPI services can plot charts and graphs and output their graphical result through `Response`.

Plotting data models

With the help of the `numpy` and `pandas` modules, FastAPI services can generate and render different types of graphs and charts using the `matplotlib` utilities. Like in the previous discussions, we will utilize an `io.ByteIO` stream and `StreamResponse` to generate graphical results for the API endpoints. The following API service retrieves survey data from the repository, computes the mean for each data strata, and returns a line graph of the data in PNG format:

```
from io import BytesIO
import matplotlib.pyplot as plt
from survey.repository.answers import AnswerRepository
from survey.repository.location import LocationRepository


@router.get("/answers/line")
async def plot_answers_mean():
    x = [1, 2, 3, 4, 5, 6, 7]
    repo_loc = LocationRepository()
    repo_answers = AnswerRepository()
    locations = await repo_loc.get_all_location()
    temp = []
    data = []
    for loc in locations:
```

```
        for qid in range(1, 13):
            loc_q1 = await repo_answers
                .get_answers_per_q(loc["id"], qid)
            if not len(loc_q1) == 0:
                loc_data = [ weights[qid-1]
                    [str(item["answer_choice"])]
                        for item in loc_q1]
                temp.append(loc_data)
        temp = list(itertools.chain(*temp))
        if not len(temp) == 0:
            data.append(temp)
        temp = list()
    y = list(map(np.mean, data))
    filtered_image = BytesIO()
    plt.figure()

    plt.plot(x, y)

    plt.xlabel('Question Mean Score')
    plt.ylabel('State/Province')
    plt.title('Linear Plot of Poverty Status')

    plt.savefig(filtered_image, format='png')
    filtered_image.seek(0)

    return StreamingResponse(filtered_image,
            media_type="image/png")
```

The plot_answers_mean() service utilizes the plot() method of the matplotlib module to plot the app's mean survey results per location on a line graph. Instead of saving the file to the filesystem, the service stores the image in the io.ByteIO stream using the module's savefig() method. The stream is rendered using StreamResponse, like in the previous samples. The following figure shows the rendered stream image in PNG format through StreamResponse:

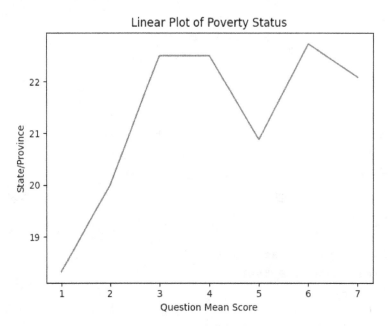

Figure 10.3 – Line graph from StreamResponse

The other API services of our app, such as `plot_sparse_data()`, create a bar chart image in JPEG format of some simulated or derived data:

```
@router.get("/sparse/bar")
async def plot_sparse_data():
    df = pd.DataFrame(np.random.randint(10, size=(10, 4)),
        columns=["Area 1", "Area 2", "Area 3", "Area 4"])
    filtered_image = BytesIO()
    plt.figure()
    df.sum().plot(kind='barh', color=['red', 'green',
            'blue', 'indigo', 'violet'])
    plt.title("Respondents in Survey Areas")
    plt.xlabel("Sample Size")
    plt.ylabel("State")
    plt.savefig(filtered_image, format='png')

    filtered_image.seek(0)
    return StreamingResponse(filtered_image,
            media_type="image/jpeg")
```

The approach is the same as our line graph rendition. With the same strategy, the following service creates a pie chart that shows the percentage of male and female respondents that were surveyed:

```
@router.get("/respondents/gender")
async def plot_pie_gender():
    repo = RespondentRepository()
    count_male = await repo.list_gender('M')
    count_female = await repo.list_gender('F')
    gender = [len(count_male), len(count_female)]
    filtered_image = BytesIO()
    my_labels = 'Male','Female'
    plt.pie(gender,labels=my_labels,autopct='%1.1f%%')
    plt.title('Gender of Respondents')
    plt.axis('equal')
    plt.savefig(filtered_image, format='png')
    filtered_image.seek(0)

    return StreamingResponse(filtered_image,
                media_type="image/png")
```

The responses generated by the plot_sparse_data() and plot_pie_gender() services are as follows:

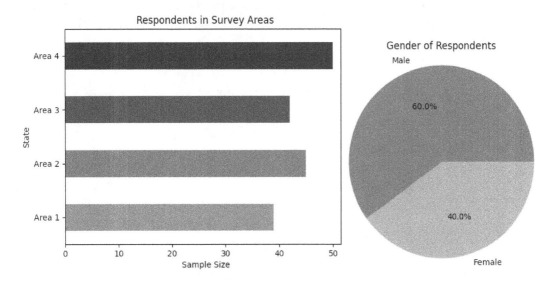

Figure 10.4 – The bar and pie charts generated by StreamResponse

This section will introduce an approach to creating API endpoints that produce graphical results using `matplotlib`. But there are other descriptive, complex, and stunning graphs and charts that you can create in less time using `numpy`, `pandas`, `matplotlib`, and the FastAPI framework. These extensions can even solve complex mathematical and data science-related problems, given the right hardware resources.

Now, let us shift our focus to the other project, `ch10-mongo`, to tackle topics regarding workflows, GraphQL, and Neo4j graph database transactions and how FastAPI can utilize them.

Simulating a BPMN workflow

Although the FastAPI framework has no built-in utilities to support its workflows, it is flexible and fluid enough to be integrated into other workflow tools such as Camunda and Apache Airflow through extension modules, middleware, and other customizations. But this section will only focus on the raw solution of simulating BPMN workflows using Celery, which can be extended to a more flexible, real-time, and enterprise-grade approach such as Airflow integration.

Designing the BPMN workflow

The `ch10-mongo` project has implemented the following BPMN workflow design using Celery:

- A sequence of service tasks that derives the percentage of the survey data result, as shown in the following diagram:

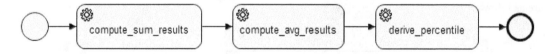

Figure 10.5 – Percentage computation workflow design

- A group of batch operations that saves data to CSV and XLSX files, as shown in the following diagram:

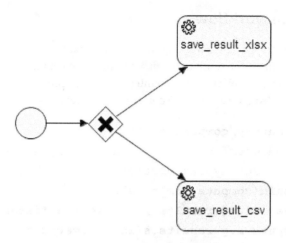

Figure 10.6 – Data archiving workflow design

- A group of chained tasks that operates on each location's data independently, as shown in the following diagram:

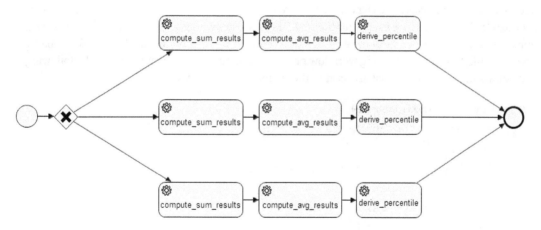

Figure 10.7 – Workflow design for stratified survey data analysis

There are many ways to implement the given design, but the most immediate solution is to utilize the Celery setup that we used in *Chapter 7, Securing the REST APIs*.

Implementing the workflow

Celery's `chain()` method implements a workflow of linked task executions, as depicted in *Figure 10.5*, where every parent task returns the result to the first parameter of next task. The chained workflow works if each task runs successfully without encountering any exceptions at runtime. The following is the API service in `/api/survey_workflow.py` that implements the chained workflow:

```python
@router.post("/survey/compute/avg")
async def chained_workflow(surveydata: SurveyDataResult):
    survey_dict = surveydata.dict(exclude_unset=True)
    result = chain(compute_sum_results
        .s(survey_dict['results']).set(queue='default'),
            compute_avg_results.s(len(survey_dict))
            .set(queue='default'), derive_percentile.s()
            .set(queue='default')).apply_async()
    return {'message' : result.get(timeout = 10) }
```

`compute_sum_results()`, `compute_avg_results()`, and `derive_percentile()` are bound tasks. Bound tasks are Celery tasks that are implemented to have the first method parameter allocated to the task instance itself, thus the `self` keyword appearing in its parameter list. Their task implementation always has the `@celery.task(bind=True)` decorator. The Celery task manager prefers bound tasks when applying workflow primitive signatures to create workflows. The following code shows the bound tasks that are used in the chained workflow design:

```python
@celery.task(bind=True)
def compute_sum_results(self, results:Dict[str, int]):
    scores = []
    for key, val in results.items():
        scores.append(val)
    return sum(scores)
```

`compute_sum_results()` computes the total survey result per state, while `compute_avg_results()` consumes the sum computed by `compute_sum_results()` to derive the mean value:

```python
@celery.task(bind=True)
def compute_avg_results(self, value, len):
    return (value/len)
```

On the other hand, `derive_percentile()` consumes the mean values produced by `compute_avg_results()` to return a percentage value:

```python
@celery.task(bind=True)
def derive_percentile(self, avg):
    percentage = f"{avg:.0%}"
    return percentage
```

The given `derive_percentile()` consumes the mean values produced by `compute_avg_results()` to return a percentage value.

To implement the gateway approach, Celery has a `group()` primitive signature, which is used to implement parallel task executions, as depicted in *Figure 10.6*. The following API shows the implementation of the workflow structure with parallel executions:

```python
@router.post("/survey/save")
async def grouped_workflow(surveydata: SurveyDataResult):
    survey_dict = surveydata.dict(exclude_unset=True)
    result = group([save_result_xlsx
        .s(survey_dict['results']).set(queue='default'),
            save_result_csv.s(len(survey_dict))
            .set(queue='default')]).apply_async()
    return {'message' : result.get(timeout = 10) }
```

The workflow shown in *Figure 10.7* depicts a mix of grouped and chained workflows. It is common for many real-world microservice applications to solve workflow-related problems with a mixture of different Celery signatures, including `chord()`, `map()`, and `starmap()`. The following script implements a workflow with mixed signatures:

```python
@router.post("/process/surveys")
async def process_surveys(surveys: List[SurveyDataResult]):
    surveys_dict = [s.dict(exclude_unset=True)
        for s in surveys]
    result = group([chain(compute_sum_results
        .s(survey['results']).set(queue='default'),
            compute_avg_results.s(len(survey['results']))
            .set(queue='default'), derive_percentile.s()
            .set(queue='default')) for survey in
                surveys_dict]).apply_async()
    return {'message': result.get(timeout = 10) }
```

The Celery signature plays an essential role in building workflows. A `signature()` method or `s()` that appears in the construct manages the execution of the task, which includes accepting the initial task parameter value(s) and utilizing the queues that the Celery worker uses to load tasks. As discussed in *Chapter 7, Securing the REST APIs*, `apply_async()` triggers the whole workflow execution and retrieves the result.

Aside from workflows, the FastAPI framework can also use the GraphQL platform to build CRUD transactions, especially when dealing with a large amount of data in a microservice architecture.

Using GraphQL queries and mutations

GraphQL is an API standard that implements REST and CRUD transactions at the same time. It is a high-performing platform that's used in building REST API endpoints that only need a few steps to set up. Its objective is to create endpoints for data manipulation and query transactions.

Setting up the GraphQL platform

Python extensions such as Strawberry, Ariadne, Tartiflette, and Graphene support GraphQL-FastAPI integration. This chapter introduces the use of the new Ariadne 3.x to build CRUD transactions for this `ch10-mongo` project with MongoDB as the repository.

First, we need to install the latest `graphene` extension using the `pip` command:

```
pip install graphene
```

Among the GraphQL libraries, Graphene is the easiest to set up, with fewer decorators and methods to override. It easily integrates with the FastAPI framework without requiring additional middleware and too much auto-wiring.

Creating the record insertion, update, and deletion

Data manipulation operations are always part of GraphQL's mutation mechanism. This is a GraphQL feature that modifies the server-side state of the application and returns arbitrary data as a sign of a successful change in the state. The following is an implementation of a GraphQL mutation that inserts, deletes, and updates records:

```
from models.data.pccs_graphql import LoginData
from graphene import String, Int, Mutation, Field
from repository.login import LoginRepository


class CreateLoginData(Mutation):
    class Arguments:
```

```
    id = Int(required=True)
    username = String(required=True)
    password = String(required=True)

ok = Boolean()
loginData = Field(lambda: LoginData)

async def mutate(root, info, id, username, password):
    login_dict = {"id": id, "username": username,
                "password": password}
    login_json = dumps(login_dict, default=json_serial)
    repo = LoginRepository()
    result = await repo.add_login(loads(login_json))
    if not result == None:
      ok = True
    else:
      ok = False
    return CreateLoginData(loginData=result, ok=ok)
```

CreateLoginData is a mutation that adds a new login record to the data store. The inner class, Arguments, indicates the record fields that will comprise the new login record for insertion. These arguments must appear in the overridden mutate() method to capture the values of these fields. This method will also call the ORM, which will persist the newly created record.

After a successful insert transaction, mutate() must return the class variables defined inside a mutation class such as ok and the loginData object. These returned values must be part of the mutation instance.

Updating a login attribute has a similar implementation to CreateLoginData except the arguments need to be exposed. The following is a mutation class that updates the password field of a login record that's been retrieved using its username:

```
class ChangeLoginPassword(Mutation):
    class Arguments:
      username = String(required=True)
      password = String(required=True)

ok = Boolean()
loginData = Field(lambda: LoginData)
```

```
async def mutate(root, info, username, password):
    repo = LoginRepository()
    result = await repo.change_password(username,
            password)

    if not result == None:
      ok = True
    else:
      ok = False
    return CreateLoginData(loginData=result, ok=ok)
```

Similarly, the delete mutation class retrieves a record through an `id` and deletes it from the data store:

```
class DeleteLoginData(Mutation):
    class Arguments:
      id = Int(required=True)

    ok = Boolean()
    loginData = Field(lambda: LoginData)

    async def mutate(root, info, id):
        repo = LoginRepository()
        result = await repo.delete_login(id)
        if not result == None:
          ok = True
        else:
          ok = False
        return DeleteLoginData(loginData=result, ok=ok)
```

Now, we can store all our mutation classes in an `ObjectType` class that exposes these transactions to the client. We assign field names to each `Field` instance of the given mutation classes. These field names will serve as the query names of the transactions. The following code shows the `ObjectType` class, which defines our `CreateLoginData`, `ChangeLoginPassword`, and `DeleteLoginData` mutations:

```
class LoginMutations(ObjectType):
    create_login = CreateLoginData.Field()
```

```
edit_login = ChangeLoginPassword.Field()
delete_login = DeleteLoginData.Field()
```

Implementing the query transactions

GraphQL query transactions are implementations of the ObjectType base class. Here, LoginQuery retrieves all login records from the data store:

```
class LoginQuery(ObjectType):
    login_list = None
    get_login = Field(List(LoginData))

    async def resolve_get_login(self, info):
        repo = LoginRepository()
        login_list = await repo.get_all_login()
        return login_list
```

The class must have a query field name, such as get_login, that will serve as its query name during query execution. The field name must be part of the resolve_*() method name for it to be registered under the ObjectType class. A class variable, such as login_list, must be declared for it to contain all the retrieved records.

Running the CRUD transactions

We need a GraphQL schema to integrate the GraphQL components and register the mutation and query classes for the FastAPI framework before running the GraphQL transactions. The following script shows the instantiation of GraphQL's Schema class with LoginQuery and LoginMutations:

```
from graphene import Schema
schema = Schema(query=LoginQuery, mutation=LoginMutations,
    auto_camelcase=False)
```

We set the auto_camelcase property of the Schema instance to False to maintain the use of the original field names with an underscore and avoid the camel case notation approach.

Afterward, we use the schema instance to create the `GraphQLApp()` instance. GraphQLApp is equivalent to an application that needs mounting to the FastAPI framework. We can use the `mount()` utility of FastAPI to integrate the `GraphQLApp()` instance with its URL pattern and the chosen GraphQL browser tool to run the API transactions. The following code shows how to integrate the GraphQL applications with Playground as the browser tool to run the APIs:

```
from starlette_graphene3 import GraphQLApp,
            make_playground_handler


app = FastAPI()
app.mount("/ch10/graphql/login",
        GraphQLApp(survey_graphene_login.schema,
            on_get=make_playground_handler()) )
app.mount("/ch10/graphql/profile",
        GraphQLApp(survey_graphene_profile.schema,
            on_get=make_playground_handler()) )
```

We can use the left-hand side panel to insert a new record through a JSON script containing the field name of the `CreateLoginData` mutation, which is `create_login`, along with passing the necessary record data, as shown in the following screenshot:

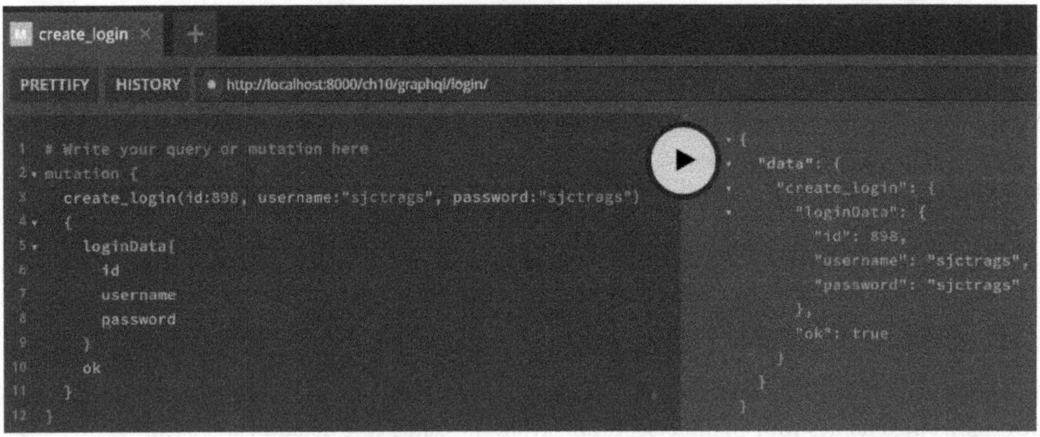

Figure 10.8 – Running the create_login mutation

To perform query transactions, we must create a JSON script with a field name of `LoginQuery`, which is `get_login`, together with the record fields needed to be retrieved. The following screenshot shows how to run the `LoginQuery` transaction:

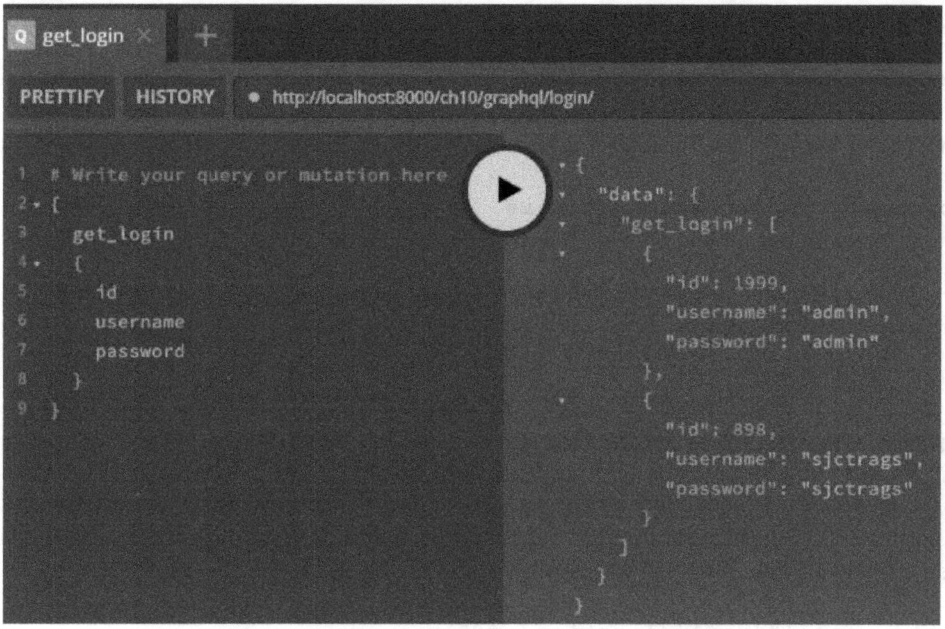

Figure 10.9 – Running the get_login query transaction

GraphQL can help consolidate all the CRUD transactions from different microservices with easy setup and configuration. It can serve as an API Gateway where all GraphQLApps from multiple microservices are mounted to create a single façade application. Now, let us integrate FastAPI into a graph database.

Utilizing the Neo4j graph database

For an application that requires storage that emphasizes relationships among data records, a graph database is an appropriate storage method to use. One of the platforms that use graph databases is Neo4j. FastAPI can easily integrate with Neo4j, but we need to install the `Neo4j` module using the `pip` command:

```
pip install neo4j
```

Neo4j is a NoSQL database with a flexible and powerful data model that can manage and connect different enterprise-related data based on related attributes. It has a semi-structured database architecture with simple ACID properties and a non-JOIN policy that make its operations fast and easy to execute.

> **Note**
> ACID, which stands for atomicity, consistency, isolation, and durability, describes a database transaction as a group of operations that performs as a single unit with correctness and consistency.

Setting the Neo4j database

The neo4j module includes neo4j-driver, which is needed to establish a connection with the graph database. It needs a URI that contains the bolt protocol, server address, and port. The default database port to use is 7687. The following script shows how to create Neo4j database connectivity:

```python
from neo4j import GraphDatabase


uri = "bolt://127.0.0.1:7687"
driver = GraphDatabase.driver(uri, auth=("neo4j",
    "admin2255"))
```

Creating the CRUD transactions

Neo4j has a declarative graph query language called Cypher that allows CRUD transactions of the graph database. These Cypher scripts need to be encoded as str SQL commands to be executed by its query runner. The following API service adds a new database record to the graph database:

```python
@router.post("/neo4j/location/add")
def create_survey_loc(node_name: str,
        node_req_atts: LocationReq):
    node_attributes_dict =
        node_req_atts.dict(exclude_unset=True)
    node_attributes = '{' + ', '.join(f'{key}:\'{value}\''
        for (key, value) in node_attributes_dict.items())
            + '}'
    query = f"CREATE ({node_name}:Location
        {node_attributes})"
    try:
        with driver.session() as session:
            session.run(query=query)
        return JSONResponse(content={"message":
            "add node location successful"}, status_code=201)
    except Exception as e:
        print(e)
        return JSONResponse(content={"message": "add node
            location unsuccessful"}, status_code=500)
```

`create_survey_loc()` adds new survey location details to the Neo4j database. A record is considered a node in the graph database with a name and attributes equivalent to the record fields in the relational databases. We use the connection object to create a session that has a `run()` method to execute Cypher scripts.

The command to add a new node is `CREATE`, while the syntax to update, delete, and retrieve nodes can be added with the `MATCH` command. The following `update_node_loc()` service searches for a particular node based on the node's name and performs the `SET` command to update the given fields:

```
@router.patch("/neo4j/update/location/{id}")
async def update_node_loc(id:int,
            node_req_atts: LocationReq):
    node_attributes_dict =
        node_req_atts.dict(exclude_unset=True)
    node_attributes = '{' + ', '.join(f'{key}:\'{value}\''
        for (key, value) in
            node_attributes_dict.items()) + '}'
    query = f"""
        MATCH (location:Location)
        WHERE ID(location) = {id}
        SET location += {node_attributes}"""
    try:
        with driver.session() as session:
            session.run(query=query)
        return JSONResponse(content={"message":
            "update location successful"}, status_code=201)
    except Exception as e:
        print(e)
        return JSONResponse(content={"message": "update
            location  unsuccessful"}, status_code=500)
```

Likewise, the delete transaction uses the `MATCH` command to search for the node to be deleted. The following service implements `Location` node deletion:

```
@router.delete("/neo4j/delete/location/{node}")
def delete_location_node(node:str):
    node_attributes = '{' + f"name:'{node}'" + '}'
    query = f"""
        MATCH (n:Location {node_attributes})
```

```
        DETACH DELETE n
"""
try:
    with driver.session() as session:
        session.run(query=query)
    return JSONResponse(content={"message":
      "delete location node successful"},
        status_code=201)
except:
    return JSONResponse(content={"message":
      "delete location node unsuccessful"},
        status_code=500)
```

When retrieving nodes, the following service retrieves all the nodes from the database:

```
@router.get("/neo4j/nodes/all")
async def list_all_nodes():
    query = f"""
        MATCH (node)
        RETURN node"""
    try:
        with driver.session() as session:
            result = session.run(query=query)
            nodes = result.data()
        return nodes
    except Exception as e:
        return JSONResponse(content={"message": "listing
            all nodes unsuccessful"}, status_code=500)
```

The following service only retrieves a single node based on the node's id:

```
@router.get("/neo4j/location/{id}")
async def get_location(id:int):
    query = f"""
        MATCH (node:Location)
        WHERE ID(node) = {id}
        RETURN node"""
    try:
```

```
        with driver.session() as session:
            result = session.run(query=query)
            nodes = result.data()
        return nodes
    except Exception as e:
        return JSONResponse(content={"message": "get
          location node unsuccessful"}, status_code=500)
```

Our implementation will not be complete if we have no API endpoint that will link nodes based on attributes. Nodes are linked to each other based on relationship names and attributes that are updatable and removable. The following API endpoint creates a node relationship between the Location nodes and Respondent nodes:

```
@router.post("/neo4j/link/respondent/loc")
def link_respondent_loc(respondent_node: str,
    loc_node: str, node_req_atts:LinkRespondentLoc):
    node_attributes_dict =
        node_req_atts.dict(exclude_unset=True)

    node_attributes = '{' + ', '.join(f'{key}:\'{value}\''
        for (key, value) in
            node_attributes_dict.items()) + '}'

    query = f"""
        MATCH (respondent:Respondent), (loc:Location)
        WHERE respondent.name = '{respondent_node}' AND
            loc.name = '{loc_node}'
        CREATE (respondent) -[relationship:LIVES_IN
            {node_attributes}]->(loc)"""
    try:
        with driver.session() as session:
            session.run(query=query)
        return JSONResponse(content={"message": "add …
            relationship successful"}, status_code=201)
    except:
        return JSONResponse(content={"message": "add
```

```
respondent-loc relationship unsuccessful"},
              status_code=500)
```

The FastAPI framework can easily integrate into any database platform. The previous chapters have proven that FastAPI can deal with relational database transactions with ORM and document-based NoSQL transactions with ODM, while this chapter has proven the same for the Neo4j graph database due to its easy configurations.

Summary

This chapter introduced the scientific side of FastAPI by showing that API services can provide numerical computation, symbolic formulation, and graphical interpretation of data via the numpy, pandas, sympy, and matplotlib modules. This chapter also helped us understand how far we can integrate FastAPI with new technology and design strategies to provide new ideas for the microservice architecture, such as using GraphQL to manage CRUD transactions and Neo4j for real-time and node-based data management. We also introduced the basic approach that FastAPI can apply to solve various BPMN workflows using Celery tasks. With this, we have started to understand the power and flexibility of the framework in building microservice applications.

The next chapter will cover the last set of topics to complete our deep dive into FastAPI. We will cover some deployment strategies, Django and Flask integrations, and other microservice design patterns that haven't been discussed in the previous chapters.

11

Adding Other Microservice Features

Our long journey of exploring FastAPI's extensibility in building microservice applications will end with this chapter, which covers standard recommendations on project setup, maintenance, and deployment using some microservice-related tools based on design patterns. This chapter will discuss the *OpenTracing* mechanism and its use in a distributed FastAPI architecture setup using tools such as *Jaeger* and `StarletteTracingMiddleWare`. The *service registry* and *client-side discovery* design patterns are included likewise in the detailed discussions on how to manage access to the API endpoints of the microservices. A microservice component that checks for the *health* of the API endpoints will also be part of the discussion. Moreover, the chapter will not end without recommendations on the FastAPI application's *deployment*, which might lead to other design strategies and network setups.

The main goal of this chapter is to complete the design architecture of a FastAPI application before its sign-off. Here are the topics that will complete our FastAPI application development venture:

- Setting up the virtual environment
- Checking the API properties
- Implementing open tracing mechanisms
- Setting up service registry and client-side service discovery
- Deploying and running applications using Docker
- Using Docker Compose for deployment
- Utilizing NGINX as an API gateway
- Integrating Django and Flask sub-applications

Technical requirements

Our last software prototype will be an **Online Sports Management System** (**OSMS**) that will manage administrators, referees, players, schedules, and game results of a tournament or league. The application will utilize MongoDB as the database storage. All of the code has been uploaded to `https://github.com/PacktPublishing/Building-Python-Microservices-with-FastAPI` under `ch11` and other *Chapter 11*-related projects.

Setting up the virtual environment

Let us start with the proper way of setting up the development environment of our FastAPI application. In Python development, it is common to manage the libraries and extension modules that are needed using a virtual environment. A virtual environment is a way of creating multiple different and parallel installations of Python interpreters and their dependencies where each has the application(s) to be compiled and run. Each instance has its own set of libraries depending on the requirements of its application(s). But first, we need to install the `virtualenv` module to pursue the creation of these instances:

```
pip install virtualenv
```

The following list describes the benefits of having a virtual environment:

- To avoid the overlapping of the library version

- To avoid broken installed module files due to namespace collisions

- To localize the libraries to avoid conflicts with the globally installed modules on which some applications are very dependent

- To create a template or baseline copy of the set of modules to be replicated on some related projects

- To maintain operating system performance and setup

After the installation, we need to run the `python -m virtualenv` command to create an instance. *Figure 11.1* shows how the `ch01-env` virtual environment for the `ch01` project is created:

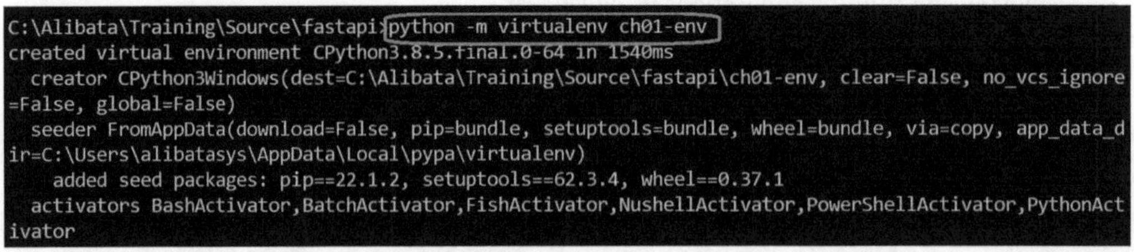

```
C:\Alibata\Training\Source\fastapi>python -m virtualenv ch01-env
created virtual environment CPython3.8.5.final.0-64 in 1540ms
  creator CPython3Windows(dest=C:\Alibata\Training\Source\fastapi\ch01-env, clear=False, no_vcs_ignore
=False, global=False)
  seeder FromAppData(download=False, pip=bundle, setuptools=bundle, wheel=bundle, via=copy, app_data_d
ir=C:\Users\alibatasys\AppData\Local\pypa\virtualenv)
    added seed packages: pip==22.1.2, setuptools==62.3.4, wheel==0.37.1
  activators BashActivator,BatchActivator,FishActivator,NushellActivator,PowerShellActivator,PythonAct
ivator
```

Figure 11.1 – Creating a Python virtual environment

To use the virtual environment, we need to configure our *VS Code editor* to utilize the Python interpreter of the virtual environment instead of the global interpreter to install modules, compile, and run the application. Pressing *Ctrl + Shift + P* will open the *Command Palette* showing the Python command to *select the interpreter*. *Figure 11.2* shows the process of choosing the Python interpreter for the ch01 project:

Figure 11.2 – Choosing the Python interpreter

The select command will open a pop-up Windows *File Explorer* window to search for the appropriate virtual environment with the Python interpreter, as shown in *Figure 11.3*:

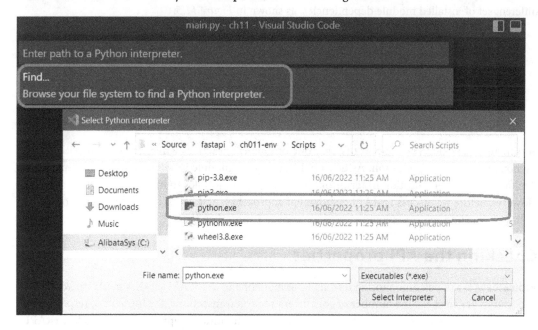

Figure 11.3 – Searching for the virtual environment

Opening a *Terminal console* for the project will automatically activate the virtual environment by running the /Scripts/activate.bat command for the Windows operating system. Additionally, this activate.bat script can be manually run if the automated activation was not successful. By the way, the activation will not be feasible with the Powershell terminal, but only with the command console, as shown in *Figure 11.4*:

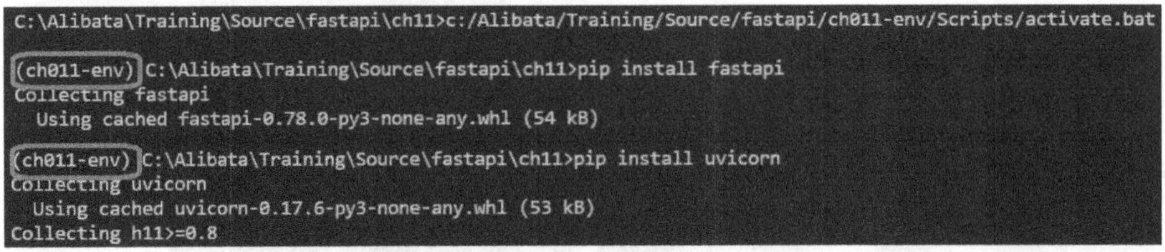

```
C:\Alibata\Training\Source\fastapi\ch11>c:/Alibata/Training/Source/fastapi/ch011-env/Scripts/activate.bat

(ch011-env) C:\Alibata\Training\Source\fastapi\ch11>pip install fastapi
Collecting fastapi
  Using cached fastapi-0.78.0-py3-none-any.whl (54 kB)
(ch011-env) C:\Alibata\Training\Source\fastapi\ch11>pip install uvicorn
Collecting uvicorn
  Using cached uvicorn-0.17.6-py3-none-any.whl (53 kB)
Collecting h11>=0.8
```

Figure 11.4 – Activating the virtual environment

After activation, we can determine the name of the activated virtual environment from the leftmost part of the command line. *Figure 11.4* shows that the Python interpreter of ch11-env is the chosen interpreter for the project. Anything installed by its pip command will only be available within that instance.

Each of our projects has a virtual environment, thus having multiple virtual environments containing different set of installed module dependencies, as shown in *Figure 11.5*:

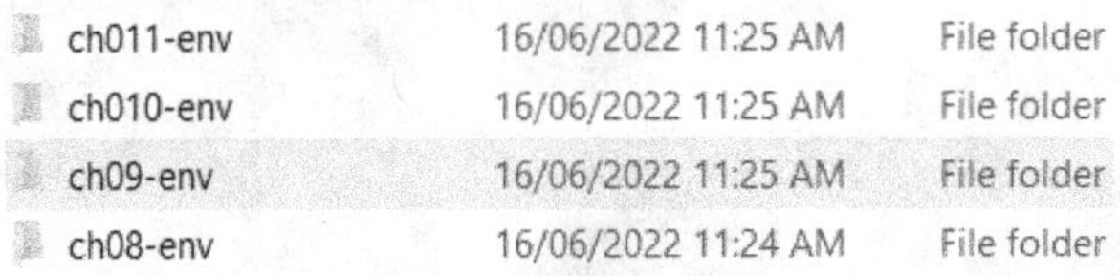

ch011-env	16/06/2022 11:25 AM	File folder
ch010-env	16/06/2022 11:25 AM	File folder
ch09-env	16/06/2022 11:25 AM	File folder
ch08-env	16/06/2022 11:24 AM	File folder

Figure 11.5 – Creating multiple virtual environments

Setting up the virtual environment is only one of the best practices when it comes to initiating a Python microservice application. Aside from localizing the module installation, it helps prepare the deployment of the application in terms of identifying what modules to install in the cloud servers. However, before we discuss FastAPI deployment approaches, first, let us discuss what microservice utilities to include before deploying a project, such as **Prometheus**.

Checking the API properties

Prometheus is a popular monitoring total that can monitor and check API services in any microservice application. It can check the number of concurrent request transactions, the number of responses at a certain period, and the total incoming requests of an endpoint. To apply Prometheus to FastAPI applications, first, we need to install the following module:

```
pip install starlette-exporter
```

Then, we add `PrometheusMiddleware` to the application and enable its endpoint to observe the API's properties at runtime. The following script shows the application setup with the Prometheus monitoring module:

```
from starlette_exporter import PrometheusMiddleware,
        handle_metrics


app = FastAPI()
app.add_middleware(PrometheusMiddleware, app_name="osms")
app.add_route("/metrics", handle_metrics)
```

Here, we add `PrometheusMiddleware` using the `add_middleware()` method of FastAPI. Then, we add an arbitrary URI pattern to the `handle_metrics()` utility to expose all of the API health details. Accessing `http://localhost:8000/metrics` will provide us with something as shown in *Figure 11.6*:

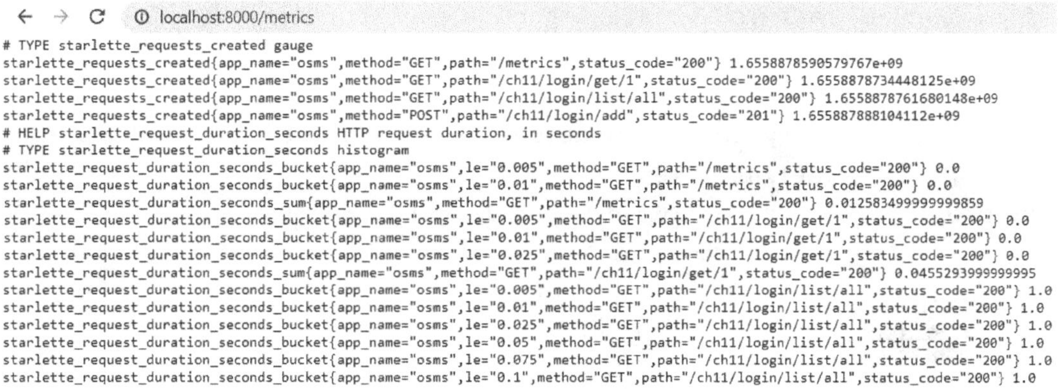

```
← → C  ⓘ localhost:8000/metrics
# TYPE starlette_requests_created gauge
starlette_requests_created{app_name="osms",method="GET",path="/metrics",status_code="200"} 1.6558878590579767e+09
starlette_requests_created{app_name="osms",method="GET",path="/ch11/login/get/1",status_code="200"} 1.6558878734448125e+09
starlette_requests_created{app_name="osms",method="GET",path="/ch11/login/list/all",status_code="200"} 1.6558878761680148e+09
starlette_requests_created{app_name="osms",method="POST",path="/ch11/login/add",status_code="201"} 1.655887888104112e+09
# HELP starlette_request_duration_seconds HTTP request duration, in seconds
# TYPE starlette_request_duration_seconds histogram
starlette_request_duration_seconds_bucket{app_name="osms",le="0.005",method="GET",path="/metrics",status_code="200"} 0.0
starlette_request_duration_seconds_bucket{app_name="osms",le="0.01",method="GET",path="/metrics",status_code="200"} 0.0
starlette_request_duration_seconds_sum{app_name="osms",method="GET",path="/metrics",status_code="200"} 0.012583499999999859
starlette_request_duration_seconds_bucket{app_name="osms",le="0.005",method="GET",path="/ch11/login/get/1",status_code="200"} 0.0
starlette_request_duration_seconds_bucket{app_name="osms",le="0.01",method="GET",path="/ch11/login/get/1",status_code="200"} 0.0
starlette_request_duration_seconds_bucket{app_name="osms",le="0.025",method="GET",path="/ch11/login/get/1",status_code="200"} 0.0
starlette_request_duration_seconds_sum{app_name="osms",method="GET",path="/ch11/login/get/1",status_code="200"} 0.0455293999999995
starlette_request_duration_seconds_bucket{app_name="osms",le="0.005",method="GET",path="/ch11/login/list/all",status_code="200"} 1.0
starlette_request_duration_seconds_bucket{app_name="osms",le="0.01",method="GET",path="/ch11/login/list/all",status_code="200"} 1.0
starlette_request_duration_seconds_bucket{app_name="osms",le="0.025",method="GET",path="/ch11/login/list/all",status_code="200"} 1.0
starlette_request_duration_seconds_bucket{app_name="osms",le="0.05",method="GET",path="/ch11/login/list/all",status_code="200"} 1.0
starlette_request_duration_seconds_bucket{app_name="osms",le="0.075",method="GET",path="/ch11/login/list/all",status_code="200"} 1.0
starlette_request_duration_seconds_bucket{app_name="osms",le="0.1",method="GET",path="/ch11/login/list/all",status_code="200"} 1.0
```

Figure 11.6 – Monitoring the endpoints

The data in *Figure 11.6* displays the time duration, in seconds, used by each API in processing requests, providing response to clients, and emitting the status code of each API transaction. Additionally, it includes some buckets that are built-in values used by the tool to create histograms. Aside from the histogram, Prometheus also allows the customization of some metrics inherent to a particular application.

Another way of monitoring a FastAPI microservice application is by adding an open tracing tool.

Implementing open tracing mechanisms

When monitoring multiple, independent, and distributed microservices, the *OpenTracing* mechanism is preferred when managing API logs and traces. Tools such as *Zipkin*, *Jaeger*, and *Skywalking* are popular distributed tracing systems that can provide the setup for trace and log collections. In this prototype, we will be using the Jaeger tool to manage the application's API traces and logs.

The current way to integrate an OpenTracing tool into FastAPI microservices is through the *OpenTelemetry* modules since the *Opentracing for Python* extension is already a deprecated module. To use Jaeger as the tracing service, OpenTelemetry has an *OpenTelemetry Jaeger Thrift Exporter* utility, which allows you to export traces to the Jaeger client applications. This exporter utility sends these traces to the configured agent using the Thrift compact protocol over UDP. But first, we need to install the following extension to utilize this exporter:

```
pip install opentelemetry-exporter-jaeger
```

Afterward, add the following configuration to the `main.py` file:

```
from opentelemetry import trace
from opentelemetry.exporter.jaeger.thrift import
        JaegerExporter
from opentelemetry.sdk.resources import SERVICE_NAME,
        Resource
from opentelemetry.sdk.trace import TracerProvider
from opentelemetry.sdk.trace.export import
        BatchSpanProcessor
from opentelemetry.instrumentation.fastapi import
        FastAPIInstrumentor
from opentelemetry.instrumentation.logging import
        LoggingInstrumentor


app = FastAPI()


resource=Resource.create(
        {SERVICE_NAME: "online-sports-tracer"})
tracer = TracerProvider(resource=resource)
trace.set_tracer_provider(tracer)


jaeger_exporter = JaegerExporter(
    # configure client / agent
    agent_host_name='localhost',
    agent_port=6831,
    # optional: configure also collector
    # collector_endpoint=
    #     'http://localhost:14268/api/traces?
```

```
    #                format=jaeger.thrift',
    # username=xxxx, # optional
    # password=xxxx, # optional
    # max_tag_value_length=None # optional
)
span_processor = BatchSpanProcessor(jaeger_exporter)
tracer.add_span_processor(span_processor)

FastAPIInstrumentor.instrument_app(app,
        tracer_provider=tracer)
LoggingInstrumentor().instrument(set_logging_format=True)
```

The first step in the preceding setup is to create a tracing service with a name using OpenTelemetry's Resource class. Then, we instantiate a tracer from the service resource. To complete the setup, we need to provide the tracer with `BatchSpanProcessor` instantiated through the `JaegerExporter` details to manage all of the traces and logs using a Jaeger client. A *trace* includes full-detailed information about the exchange of requests and responses among all API services and other components across the distributed setup. This is unlike a *log*, which only contains the details regarding a transaction within an application.

After the completed Jaeger tracer setup, we integrate the `tracer` client with FastAPI through `FastAPIInstrumentor`. To utilize this class, first, we need to install the following extension:

```
pip install opentelemetry-instrumentation-fastapi
```

Before we can run our application, first, we need to download a Jaeger client from `https://www.jaegertracing.io/download/`, unzip the `jaeger-xxxx-windows-amd64.tar.gz` file, and run `jaeger-all-in-one.exe`. Installers for Linux and macOS are also available.

Now, open a browser and access the Jaeger client through the default `http://localhost:16686`. *Figure 11.7* shows a snapshot of the tracer client:

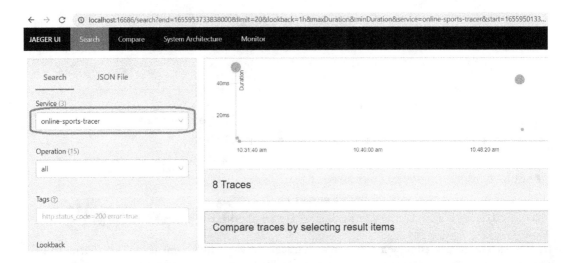

Figure 11.7 – Monitoring microservices through a Jaeger client

After some browser reloads, the Jaeger app will detect our tracer through its service name, online-sports-tracer, after running our microservice application. All accessed API endpoints are detected and monitored, thus creating traces and visual analyses regarding all requests and response transactions incurred by these endpoints. *Figure 11.8* shows the traces and graphical plots generated by Jaeger:

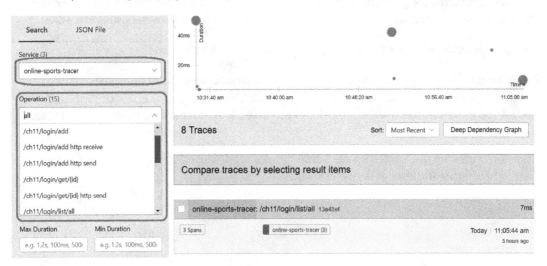

Figure 11.8 – Searching the traces of every API transaction

A span in OpenTelemetry is equivalent to a trace with a unique *ID*, and we can scrutinize each span to view all the details by clicking on the search traces for every endpoint. Clicking on the searched trace for the `/ch11/login/list/all` endpoint, as shown in *Figure 11.8*, can provide us with the following trace details:

Figure 11.9 – Scrutinizing the trace details of an endpoint

Aside from the traces shown in *Figure 11.9*, the Jaeger client can also collect the *uvicorn logs* through an OpenTelemetry module called `opentelemetry-instrumentation-logging`. After installing the module, we can enable the integration by instantiating `LoggingInstrumentor` in the `main.py` file, as shown in the previous code snippet.

Now, let us add the *service registry* and *client-side service discovery* mechanisms to our application.

Setting up service registry and client-side service discovery

A service registry tool such as *Netflix Eureka* enables the registration of microservice applications without knowing the exact DNS locations of their servers. It manages all access to these registered services using a load-balancing algorithm and dynamically assigns these service instances with network locations. This service registration is helpful to microservice applications deployed to servers with changing DNS names due to failures, upgrades, and enhancements.

For the service registry to work, the service instances should have a mechanism to discover the registry server before the server registration. For FastAPI, we need to utilize the `py_eureka_client` module to implement the service discovery design pattern.

Implementing client-side service discovery

Creating a FastAPI microservice application to discover and register to a service registry server such as *Netflix Eureka* is straightforward. First, we need to install py_eureka_client through pip:

```
pip install py_eureka_client
```

Then, we instantiate its EurekaClient component class with the correct eureka_server, app_name, instance_port, and instance_host parameter details. The eureka_server parameter must be the exact machine address of the Eureka server and not localhost. Additionally, the client instance must have the appropriate app_name parameter for the FastAPI microservice application (or client app), with the instance_port parameter set to 8000 and the instance_host to 192.XXX.XXX.XXX (not localhost or 127.0.0.1). The following snippet depicts the location in main.py in which to instantiate the EurekaClient component class:

```python
from py_eureka_client.eureka_client import EurekaClient

app = FastAPI()

@app.on_event("startup")
async def init():
    create_async_db()
    global client
    client = EurekaClient(
    eureka_server="http://DESKTOP-56HNGC9:8761/eureka",
    app_name="sports_service", instance_port=8000,
    instance_host="192.XXX.XXX.XXX")
    await client.start()

@app.on_event("shutdown")
async def destroy():
    close_async_db()
    await client.stop()
```

The client discovery happens in the `startup` event of the application. It starts with the instantiation of the `EurekaClient` component class and invoking its `start()` method either asynchronously or not. The `EurekaClient` component class can handle asynchronous or synchronous FastAPI startup events. To close the server discovery process, always invoke `Eureka Client`'s `stop()` method in the `shutdown` event. Now, let us build our Netflix Eureka server registry before running and performing the client-side service discovery.

Setting up the Netflix Eureka service registry

Let us utilize the Spring Boot platform to create our Eureka server. We can create an application through `https://start.spring.io/` or the *Spring STS IDE*, using either a Maven- or Gradle-driven application. Ours is a Maven application with `pom.xml` that has the following dependency for the Eureka Server setup:

```
<dependency>
        <groupId>org.springframework.cloud</groupId>
        <artifactId>
          spring-cloud-starter-netflix-eureka-server
        </artifactId>
</dependency>
```

In this case, `application.properties` must have `server.port` set to `8761`, `server.shutdown` enabled for a `graceful` server shutdown, and a `spring.cloud.inetutils.timeout-seconds` property set to `10` for its hostname calculation.

Now, run the Eureka Server application before the FastAPI client application. The Eureka server's logs will show us the automatic detection and registration of FastAPI's `EurekaClient`, as shown in *Figure 11.10*:

```
- [nio-8761-exec-4] c.n.e.registry.AbstractInstanceRegistry  : Registered instance UNKNOWN/DESKTOP-56HNGC9.local:8761 with status UP (replication=true)
- [nio-8761-exec-5] c.n.e.registry.AbstractInstanceRegistry  : Registered instance SPORTS_SERVICE/192.168.1.5:sports_service:8000 with status UP (replication=false)
- [nio-8761-exec-9] c.n.e.registry.AbstractInstanceRegistry  : Registered instance SPORTS_SERVICE/192.168.1.5:sports_service:8000 with status UP (replication=true)
```

Figure 11.10 – Discovering the FastAPI microservice application

The result of the client-side service discovery is also evident on the Eureka server's dashboard at `http://localhost:8761`. The page will show us all the services that consist of the registry and through which we can access and test each service. *Figure 11.11* shows a sample snapshot of the dashboard:

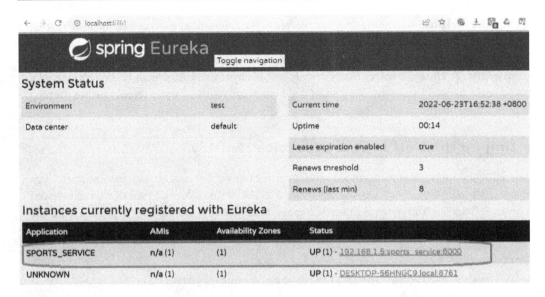

Figure 11.11 – Creating the service registry

Our **SPORTS_SERVICE** being part of the Eureka server registry, as depicted in *Figure 11.11*, means we successfully implemented the *client-side service discovery design pattern*, and it is time to deploy our application to a Docker container.

Deploying and running applications using Docker

Dockerization is a process of packaging, deploying, and running applications using Docker containers. Containerizing FastAPI microservices saves installation and setup time, space, and resources. And containerized apps are replaceable, replicable, efficient, and scalable compared to the usual deployment packaging.

To pursue Dockerization, we need to install *Docker Hub* and/or *Docker Engine* for the CLI commands. But be aware of the new Docker Desktop License Agreement (https://www.docker.com/legal/docker-software-end-user-license-agreement/) regarding its new subscription model. This chapter mainly focuses on how to run CLI commands rather than the Docker Hub GUI tool. Now, let us generate the list of modules to be installed in the docker image.

Generating the requirements.txt file

Since we are using a virtual environment instance for module management, it is easy to identify what extension modules to install in the Docker image. We can run the following command to generate a complete list of modules and their versions to the requirements.txt file:

```
pip freeze > requirements.txt
```

Then, we can create a command to copy this file to the image through the Dockerfile.

Creating the Docker image

The next step is to build a container image from any available Linux-based container images in *Docker Hub*. But we need a `Dockerfile` containing all the commands associated with pulling an available Python image from Docker Hub, creating a working directory, and copying project files from the local directory. The following is a `Dockerfile` set of instructions we use to deploy our prototype to a Python image:

```
FROM python:3.9

WORKDIR /code

COPY ./requirements.txt /code/requirements.txt
RUN pip install --no-cache-dir --upgrade -r
              /code/requirements.txt
COPY ./ch11 /code
EXPOSE 8000
CMD ["uvicorn", "main:app", "--host=0.0.0.0" , "--reload" ,
    "--port", "8000"]
```

The first line is an instruction that will derive a Python image, usually Linux-based, with an installed Python 3.9 interpreter. The command after that creates an arbitrary folder, `/code`, which will become the application's main folder. The `COPY` command copies our `requirements.txt` file to the `/code` folder, and then the `RUN` instruction installs the updated modules from the `requirements.txt` list using the following command:

```
pip install -r requirements.txt
```

Afterward, the second `COPY` command copies our `ch11` application to the working directory. The `EXPOSE` command binds port `8000` to the local machine's port `8000` to run the `CMD` command, which is the last instruction of the `Dockerfile`. The `CMD` instruction uses *uvicorn* to run the application at port 8000 using host `0.0.0.0` and not `localhost` to automatically map and utilize the IP address assigned to the image.

The `Dockerfile` must be in the same folder as the `requirements.txt` file and the `ch11` application. *Figure 11.12* shows the organization of the files and folders that needed to be Dockerized to a Python container image:

Figure 11.12 – Setting up the Docker folder structure

Once all the files and folders are complete, we run the following CLI command within the folder using the terminal console:

```
docker build -t ch11-app .
```

To check the image, run the docker image ls CLI command.

Using the Mongo Docker image

The backend of our application is MongoDB, so we need to pull the latest mongo image from Docker Hub using the following CLI command:

```
docker pull mongo:latest
```

And before we run both the ch11-app application and the mongo:latest images, first, we need to create a ch11-network by running the following command:

```
docker network create ch11-network
```

This network becomes a bridge between mongo and ch11-app once they are deployed as containers. It will establish the connectivity between the two containers to pursue the *Motor-ODM* transactions.

Creating the containers

A **container** is a running instance of a container image. We use the docker run command to start and run a pulled or created image. So, running the Mongo image using the ch11-network routes requires the execution of the following CLI command:

```
docker run --name=mongo --rm -p 27017:27017
-d                      --network=ch11-network mongo
```

Inspect the mongo:latest container using the docker inspect command to derive and use its IP address for Motor-ODM's connectivity. Replace the localhost used in AsyncIOMotorClient, which is found in the config/db.py module of ch11-app with the "inspected" IP address. Be sure to re-build the ch11-app Docker image after the update.

Now, run the `ch11-app` image with `ch11-network` using the following command:

```
docker run --name=ch11-app --rm -p 8000:8000-
d               --network=ch11-network ch11-app
```

Access the application through `http://localhost:8000/docs` to check all the API endpoints from the OpenAPI documentation.

Now, another approach to simplifying containerization is to use the *Docker Compose* tool.

Using Docker Compose for deployment

However, you need to install the Docker Compose utility in your operating system, which requires Docker Engine as the pre-installation requirement. After the installation, the next step is to create the `docker-decompose.yaml` file containing all the services needed to build the images, process the Dockerfile, build the Docker network, and create and run the containers. The following snippet shows the content of our configuration file that sets up the `mongo` and `ch11-app` containers:

```yaml
version: "3"
services:
    ch11-mongo:
        image: "mongo"
        ports:
            - 27017:27017
        expose:
            - 27017
        networks:
            - ch11-network


    ch11-app:
        build: .        # requires the Dockerfile
        depends_on:
            - ch11-mongo
        ports:
            - 8000:8000
        networks:
            - ch11-network
networks:
    ch11-network:
        driver: bridge
```

Instead of running separate Docker CLI commands, Docker Compose creates services, such as ch11-mongo and ch11-app, to manage the containerization and only uses one CLI command to execute these services, docker-compose up. The command not only creates the network of images but also runs all the containers.

One advantage of using Docker Compose is the ease of ORM and ODM configuration. Instead of performing a container inspection to understand which IP address to use, we can use the *service name of the database setup* as the hostname to establish database connectivity. It is convenient since the IP address of the mongo container varies for every instance created. The following is the new AsyncIOMotorClient with the ch11-mongo service as the hostname:

```
def create_async_db():
    global client
    client = AsyncIOMotorClient(str("ch11-mongo:27017"))
```

Now, let us implement an API Gateway design pattern for the containerized applications using the *NGINX* utility.

Using NGINX as an API Gateway

In *Chapter 4, Building the Microservice Application*, we implemented the API Gateway design pattern using only some FastAPI components. In this last chapter, we will build a *reverse proxy server* through NGINX that will assign a proxy IP address to each containerized microservice application. These proxy IPs will redirect client requests to the actual microservices running on their respective containers.

Instead of building an actual NGINX environment, we will be pulling an available NGINX image from Docker Hub to implement the reverse proxy server. This image creation requires a new Docker app folder with a different Dockerfile containing the following instructions:

```
FROM nginx:latest
COPY ./nginx_config.conf /etc/nginx/conf.d/default.conf
```

The Dockerfile instructs the creation of the latest *NGINX* image and a copy of a nginx_config.conf file to that image. The file is an *NGINX configuration file* that contains the mapping of a proxy IP address to the actual container address of each microservice application. It also exposes 8080 as its official port. The following is the content of our nginx_config.conf file:

```
server {
    listen 8080;

    location / {
        proxy_pass http://192.168.1.7:8000;
```

```
        }
    }
```

The application's OpenAPI documentation can now be accessed through `http://localhost:8080/docs`.

The Dockerization of NGINX must come after deploying applications to the containers. But another approach is to include NGINX's `Dockerfile` instructions in the application's `Dockerfile` to save time and effort. Or we can create another service in the `docker-decompose.yaml` file to build and run the NGINX image.

And for the last time, let us explore the power of FastAPI in its integration with other popular Python frameworks such as *Flask* and *Django*.

Integrating Flask and Django sub-applications

Flask is a lightweight framework that is popular for its *Jinja2* templates and *WSGI* server. On the other hand, *Django* is a Python framework that promotes rapid development using CLI commands and applies the scaffolding of files and folders to build projects and applications. Django applications can run on either WSGI- or ASGI-based servers.

We can create, deploy, and run Flask and Django projects inside a FastAPI microservice application. The framework has `WSGIMiddleware` to wrap both Flask and Django applications and integrate them into the FastAPI platform. Running the FastAPI application through *uvicorn* will also run both applications.

Of the two, it is easier to integrate the Flask application with a project than Django. We only need to import the Flask `app` object into the `main.py` file, wrap it with `WSGIMiddleware`, and mount it into the FastAPI app object. The following script shows the part of `main.py` that integrates our `ch11_flask` project:

```
from ch11_flask.app import app as flask_app
from fastapi.middleware.wsgi import WSGIMiddleware
app.mount("/ch11/flask", WSGIMiddleware(flask_app))
```

All API endpoints implemented in `ch11_flask` will be accessed using the URL prefix, `/ch11/flask`, as indicated in the `mount()` method. *Figure 11.13* shows the location of `ch11_flask` inside the `ch11` project:

Figure 11.13 – Creating a Flask application inside the FastAPI project

On the other hand, the following `main.py` script integrates our `ch11_django` application into the `ch11` project:

```
import os
from django.core.wsgi import get_wsgi_application
from importlib.util import find_spec
from fastapi.staticfiles import StaticFiles


os.environ.setdefault('DJANGO_SETTINGS_MODULE',
            'ch11_django.settings')
django_app = get_wsgi_application()

app = FastAPI()
app.mount('/static',
    StaticFiles(
        directory=os.path.normpath(
```

```
            os.path.join(
        find_spec('django.contrib.admin').origin,
                '..', 'static')
        )
    ),
    name='static',
)
app.mount('/ch11/django', WSGIMiddleware(django_app))
```

The Django framework has a `get_wsgi_application()` method that is uses to retrieve its app instance. This instance needs to be wrapped by `WSGIMiddleware` and mounted into the FastAPI app object. Moreover, we need to load the `settings.py` module of the `ch11_django` project into the FastAPI platform for global access. Also, we need to mount all the static files of the `django.contrib.main` module, which includes some HTML templates of the Django *security module.*

All views and endpoints created by the `sports` application of the `ch11_django` project must be accessed using the `/ch11/django` URL prefix. *Figure 11.14* shows the placement of the `ch11_django` project within the ch11 app:

Figure 11.14 – Creating a Django project and application inside a FastAPI object

Summary

The last chapter has given us the avenue on how to start, deploy, and run a FastAPI microservice application that follows the standards and best practices. It introduces the use of a virtual environment instance to control and manage the installation of modules from the start of the development until the deployment of our applications to Docker containers. The chapter has extensively explained the approaches on how to package, deploy, and run containerized applications. And lastly, the chapter has implemented an NGINX reverse proxy server for the application to build the API Gateway for our specimen.

Right from the start, we have witnessed the simplicity, power, adaptability, and scalability of the FastAPI framework, from creating background processes to rendering data using HTML templates. Its fast execution of API endpoints through its coroutines gives the framework the edge to become one of the most popular Python frameworks in the future. As the community of FastAPI continues to grow, we hope for more promising features in its future updates, such as support for reactive programming, circuit breakers, and a signature security module. We're hoping for the best for the FastAPI framework!

Index

H

HTTP/2 protocol
  using 261
httpx module
  about 91
  using 91, 92

I

Integrated Development Environment (IDE) 4
Inversion of Control (IoC) principle
  applying 48

J

JSON-compatible types
  objects, converting to 38, 39
JSON Object Signing and
      Encryption (JOSE) 228
JWT tokens
  access_token, creating 228
  applying 228
  login transaction, creating 229, 230
  secret key, generating 228
  secured endpoints, accessing 230, 231

K

Kafka broker
  running 271
Keycloak 240
Keycloak client 241
Keycloak realm 240

L

laboratory information management
      systems (LIMS) 325
Lagom module
  container 74
  FastAPI integration 74
  using 74
layers
  creating 94
linear, and non-linear inequalities
  solving 333, 334
linear expressions
  solving 332, 333
logging mechanism
  centralizing 86
logging middleware
  building 88-90
Loguru module
  utilizing 86-88

M

main.py file
  implementing 29, 30
message-driven transactions
  building, with RabbitMQ 269
microservice configuration
  managing 100
microservice ID
  evaluating 83, 84
middleware
  about 295
  applying, to filter path operations 45, 46
mocking 323
model layer 63

W

WebSocket client
  implementing 276, 277

X

XLSX report
  generating 338-342

Z

Zookeeper server
  running 271

Other Books You May Enjoy

If you enjoyed this book, you may be interested in these other books by Packt:

Building Python Web APIs with FastAPI

Abdulazeez Abdulazeez Adeshina

ISBN: 978-1-80107-663-0

- Set up a FastAPI application that is fully functional and secure
- Perform CRUD operations using SQL and FastAPI
- Manage concurrency in FastAPI applications
- Implement authentication in a FastAPI application
- Deploy a FastAPI application to any platform

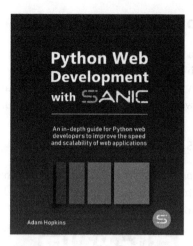

Python Web Development with Sanic

Adam Hopkins

ISBN: 978-1-80181-441-6

- Understand the difference between WSGI, Async, and ASGI servers

- Discover how Sanic organizes incoming data, why it does it, and how to make the most of it

- Implement best practices for building reliable, performant, and secure web apps

- Explore useful techniques for successfully testing and deploying a Sanic web app

- Create effective solutions for the modern web, including task management, bot integration, and GraphQL

Packt is searching for authors like you

If you're interested in becoming an author for Packt, please visit `authors.packtpub.com` and apply today. We have worked with thousands of developers and tech professionals, just like you, to help them share their insight with the global tech community. You can make a general application, apply for a specific hot topic that we are recruiting an author for, or submit your own idea.

Share your thoughts

Now you've finished *Building Python Microservices with FastAPI*, we'd love to hear your thoughts! Scan the QR code below to go straight to the Amazon review page for this book and share your feedback or leave a review on the site that you purchased it from.

https://packt.link/r/1803245964

Your review is important to us and the tech community and will help us make sure we're delivering excellent quality content.

www.ingramcontent.com/pod-product-compliance
Lightning Source LLC
Chambersburg PA
CBHW082116070326
40690CB00049B/2900